T0202003

From Mountains to Mangroves

Protecting Pakistan's Natural Heritage

From Mountains to Mangroves

Protecting Pakistan's Natural Heritage

Rina Saeed Khan

OXFORD
UNIVERSITY PRESS

Oxford University Press is a department of the University of Oxford.
It furthers the University's objective of excellence in research, scholarship,
and education by publishing worldwide. Oxford is a registered trade mark of
Oxford University Press in the UK and in certain other countries

Published in Pakistan by
Ameena Saiyid, Oxford University Press
No.38, Sector 15, Korangi Industrial Area,
PO Box 8214, Karachi-74900, Pakistan

ISBN 978-0-19-940546-6

Printed on 80gsm Local Offset Paper

Printed by Kagzi Printers, Karachi

Acknowledgements
Photography Credits: Hussain Bux Bhagat figure 11.5;
Jameel Ahmed figure 11.4; Rina Saeed Khan figures 2.1, 3.2–3.5,
4.1–4.3, 5.1–5.2, 7.1–7.2, 8.1–8.2, 9.1–9.2, 10.1–10.2, 11.1–11.3,
12.1–12.2, 13.1, 14.1–14.2, 15.1, 16.1;
Saleemullah figure 4.4; Sohail Siddiqui figure 4.5;
Syed Harir Shah figure 3.1; WWF-Pakistan figures 9.3, 15.2–15.3

Dedicated to my late father
Mohd. Saeed Khan Satti
who grew up in the forested mountains of Kotli Sattian,
where his father Khan Saheb Mohd. Azim Khan
was the last chief of the Satti tribe. From an early
age my father instilled in me a love for nature
and respect for the environment.

Contents

List of Figures

Foreword

FROM MOUNTAINS TO MANGROVES: PROTECTING PAKISTAN'S Natural Heritage is an endeavour to describe the richness and wealth of Pakistan's natural resources, the threats they face, and the many different paths to sustainable development being pursued by individuals, communities, and NGOs.

Many of the projects described in the book in vivid detail by environmental journalist Rina Saeed Khan have, over the years, been supported by WWF-Pakistan, and led to the book being supported by WWF's Small Grants Programme.

The accounts related have been selected from Rina's many travels over the past two decades to far-flung areas ranging from the Karakoram Mountains to the Makran Coast describing some of the most remarkable natural resource areas in Pakistan and outlining the significant conservation work being undertaken there. The narrative captures the rich diversity of Pakistan's unique geography, highlighting the threatened habitats of some of the rarest species on the planet as well as the people and communities endeavouring to save them.

Over the years, a growing population and increasing infrastructure needs have led to the decimation of animal habitats and forests and resulted in endangering or threatening rare species and flora and fauna. The stories collected in the book show that Pakistan has been putting in dedicated efforts to address its conservation and environment issues.

This book, which covers the survival of threatened species such as the green turtle, the Indus blind dolphin, and the straight-horned Suleiman markhor; and also focuses on protected areas in Pakistan such as the Patriata forest, Ayubia National Park, and Khunjerab National Park, is an important record both of Pakistan's surviving natural bounty and of recent conservation history.

Rina also travelled to the remote valley of Palas in Kohistan, the Kalash valleys in Chitral, and the chilgoza carpeted mountains of Zhob to highlight important work being done to preserve forests. In Sindh, she travelled to Keenjhar Lake near Karachi and Chotiari Reserve in Sanghar, both significant waterbodies that were selected as part of the Pakistan Wetlands Programme which was implemented by WWF-Pakistan. WWF-Pakistan takes enormous pleasure in supporting this book which provides us all with encouraging accounts of leadership in the quest for sustainable development.

Dr Kauser Abdulla Malik
Chairman Scientific Committee
WWF-Pakistan

Preface

MY FATHER, A RETIRED FIGHTER PILOT AND AN AVID HUNTER
in his youth, had travelled throughout Pakistan. He would
often describe those distant, romantic places to me but they
invariably sounded ever so remote and inaccessible. It was
only when I began working for *The Friday Times* (*TFT*), an
independent English language weekly newspaper based
in Lahore, as a writer and editor (after having completed
my studies in a college in the US), that my own voyage of
discovery began. What a remarkable country Pakistan has
proven to be. I often repeat this to friends, colleagues, and just
about anyone who shows any curiosity in what I write about:
how many of us are aware that Pakistan has the widest range
in land elevation of any country on earth, ranging from 8,611
m (the summit of K-2) all the way down to 0 m at sea level
(Sonmiani Bay)? The dramatic changes in altitude implies
extremely diverse physical environments with wildlife, ranging
from snow leopards in the high mountains to marine turtles
on the coast. I was first told this by Richard Garstang who
served as conservation advisor with the World Wide Fund for
Nature, Pakistan (WWF-Pakistan) for over a decade before he
retired; it was his mantra and now it has become mine.

My first column on the environment for *TFT* was on
the blind dolphins of Sindh. That was the time when my
commitment to the environment was forged during a magical

ride on a wooden boat on the Indus River near Sukkur in the mid-1990s. It was during that early morning on the mighty Indus that I began realizing what it truly meant to care for the environment. The mist had begun lifting by the time our boat was pushed out into the muddy, swirling river (so wide that the other bank, near Sukkur, was barely visible). We gazed around in hushed silence, almost mesmerized, at the ruins of old forts and crumbling shrines that remain on several of the mystical islands on the river. When our boat, a convenient wooden contraption with oars, quietly hit the rapids in the middle of the river, I could clearly hear the cries of the dolphins although it was initially difficult to spot them. With their purplish-grey colouring, they were barely discernible in the muddy waters but then, right before our eyes, a couple of them suddenly leapt up and dived right back into the river forming a graceful arc. That was truly a sight to behold.

At that instant the Fokker flight to Sindh, the Spartan rest house, the heat and stench of Sukkur, all seemed eminently worthwhile. Indeed, all else appeared completely irrelevant, especially when I learnt that this particular stretch of the Indus near the Sukkur Barrage is home to the last remaining schools of the Indus dolphin. It was tragic to think that these dolphins might not survive for very long with only around 600 individual Indus dolphins then in the river (according to WWF-Pakistan). What could I do to help these fragile creatures, as old as the river itself, survive? I could write about them and relate to people their significance and perhaps in that way play a minor role in helping them live on. There were, however, other endangered species too, such as the Suleiman markhor, snow leopard, Marco Polo sheep, green turtle, and so many others, all indigenous to Pakistan. As the eminent biologist George B. Schaller writes in his book, *Stones of Silence: Journeys in the Himalaya* (1981), about markhors: 'to think that any can have a future may be ascribed to the pathology of human idealism, but in the final analysis there is little else for which to fight'.[1]

I began my travels across Pakistan as an environmental journalist and discovered a world that I would never have known had I not been aboard that boat that morning. When I look back at all my journeys in Pakistan, they all seem to merge into a kaleidoscope of images, some too wondrous to even attempt to describe. Many people ask me, 'Which was your most amazing trip?' and I wonder! Was it that misty morning boat ride on the Indus River? Or that winding road leading up to the Khunjerab Pass, neck craning out of the window as I attempted to glimpse a herd of ibex across the ravine? Perhaps it was during that near vertical climb to the top of the cliffs of the Torghar Hills on the border with Afghanistan and my first sighting of the Suleiman markhor in the wild ...

I really cannot say for certain because for me the mountains, the deserts, even the river plains have their own unique charm. Pakistan is truly blessed with its diverse physical environments, home to unique wildlife and people living a traditional lifestyle. In an increasingly globalized world, in which sprawling megacities and corporations are homogenizing our lives, it is so refreshing and revitalizing to come into contact with people who still live in rhythm with nature, just as our ancestors once did. Although they undoubtedly endure lives of hardship, I am continually overawed by their existence in such close harmony with their natural environment. It is this very diversity that environmentalists throughout the world are striving to conserve. The world would be a far, far poorer place without it.

My adventures have been so memorable and educative that I feel they should now in this book be shared with a wider

audience to entertain and inform. The first chapter opens
with the Khunjerab National Park on the border with China
where ibex and Marco Polo sheep are found. I then move to
the Kalash valleys in Chitral where a unique indigenous tribe
lives in harmony with nature. Near Mastuj, a beautiful green
valley in Chitral, lies the Shandur Pass where the highest polo
tournament in the world is held annually, drawing thousands
of people who endanger its wilderness. Next, I journey to
Kohistan, to the remote valley of Palas (also one of the most
beautiful valleys in Pakistan), with pristine forests where
the near extinct western tragopan pheasant species resides.
Then I come down to Patriata, the highest mountain in the
Punjab, which was saved from 'development' thanks to the
efforts and lobbying by environmentalists, and then on to
Ayubia National Park in Khyber Pakhtunkhwa province where
the common leopard lives in a protected forest surrounded
by villages. I then journey to Balochistan where I visit the
Chilgoza forests of Zhob and the ancient Juniper forests of
Ziarat before heading to the border with Afghanistan, to
Torghar, where I encounter the endangered Suleiman markhor
in the wild. From there I journey deep into the Salt Range
where migratory birds come from as far afield as Siberia to
winter on the picturesque Ucchali and Khabeki lakes.

The last few chapters of the book focus on Sindh province.
There I visit the captivating Indus Dolphin Reserve in
Sukkur and the remote Chotiari Reserve in Sanghar where
one encounters the threatened marsh crocodiles and other
migratory birds. The last few chapters are on the ecology of
Keenjhar Lake, which supplies the city of Karachi with all
its freshwater requirements, and the toxic Manchar Lake,
the largest lake in Pakistan. I then journey to the coastline
of Pakistan where I visit the rehabilitation of the Mangrove

forests in Keti Bunder by WWF-Pakistan. The last chapter is
about my trip to the Makran Coast where I visit the secluded
beach in Jiwani where green turtles in large numbers come
to hatch their eggs.

In a world threatened by overpopulation and climate change,
where it is estimated that wildlife populations have dropped
sharply by 52 per cent between 1970 and 2010 (according to
WWF's recent *Living Planet Report*),[2] it becomes even more
imperative that we in Pakistan do what we can to save our part
of the planet. According to scientist Paul Ehrlich, professor
of Biological Studies at Stanford University, who has been
studying and writing about extinction for years, the 'human
population growing in numbers, per capita consumption,
and economic inequity has altered or destroyed natural
habitats. The long list of impacts include: land clearing for
farming, logging, and settlement; introduction of invasive
species; carbon emissions that drive climate change and ocean
acidification; and toxins that alter and poison ecosystems.'[3]
He says that we are sawing off the limb that we are
sitting on. Given its rapid population growth, Pakistan is
steadily depleting its natural resources more rapidly than
most countries.

However, as this book shows, there are remarkable
individuals, communities, and NGOs in Pakistan who are
demonstrating that notwithstanding a rapidly eroding
resource base, solutions can be found to address the needs
of people today while at the same time safeguarding those
of future generations which is essentially what sustainable
development is all about. They are doing their utmost to halt
and reverse the destruction of our natural landscapes and
their stories need to be shared so that others can be inspired
by their efforts. As the legendary naturalist and documentary

film-maker, Sir David Attenborough, recently put it, 'I think what's required is an understanding and a gut feeling that the natural world is part of your inheritance. This is the only planet we've got and we've got to protect it. And people do feel that, deeply and instinctively ...'.[4]

Rina Saeed Khan
Islamabad, 2016

Notes

1. George B. Schaller, *Stones of Silence: Journeys in the Himalaya* (New York: Viking Press, 1980).
2. *Living Planet Report* (WWF, 2014).
3. Rob Jordan, 'Stanford researcher declares that the sixth mass extinction is here', *Stanford News* (19 June 2015). Website: <http://news.stanford.edu/2015/06/19/mass-extinction-ehrlich-061915/>
4. Esther Addley, 'David Attenborough and Barack Obama face-to-face in a TV interview', *Guardian* (25 June 2015). Website: <https://www.theguardian.com/tv-and-radio/2015/jun/25/sir-david-attenborough-and-barack-obama-face-to-face-in-tv-interview>

1

Valley of Blood: Pakistan-China Border

MY JOURNEY TO KHUNJERAB NATIONAL PARK, LOCATED
high up in the Karakoram Mountains on the border with
China, was an adventure in itself. Mizho Ishii, a friend from
Japan, accompanied me on our drive along the Karakoram
Highway (KKH) from Gilgit airport to Karimabad in Hunza
Valley which was spectacular. Little did we know that the
best was yet to come. From Karimabad onwards to Sost, the
last border town before the rugged wilderness of Khunjerab
begins, the landscape on both sides of the KKH becomes
increasingly dramatic. The KKH is also known as the Eighth
Wonder of the World. Built jointly by China and Pakistan, the
road winds 800 memorable miles, from near the capital city
of Islamabad up to the Chinese town of Kashgar on the other
side of the Khunjerab Pass. The highway took around twenty
years to build and opened up in 1978; it is said that hundreds
of Pakistanis and Chinese died during its construction, i.e.
around one life for every kilometre.[1]

Our journey was undertaken just prior to 9/11, and there
were any number of international tourists cycling, hiking,
and trekking across this region. In Karimabad, the capital
of Hunza, I discovered a hotel occupied exclusively by the
Japanese in the summer months. My friend Mizho explained
that Hunza was an extremely popular tourist destination for
the Japanese. In fact, it even features in one of their cartoons

1

or manga (the Japanese word for the comic books that are widely read in Japan). She was extremely excited to be able to visit Hunza herself, and along the way we were able to give a lift to a lone female Japanese hitchhiker who was on her way to Lake Borith, near Gulmit in upper Hunza. We later nicknamed the waterbody Lake Bore-ith because it evoked boredom; given its sulphuric waters, one could neither wade in it nor cross it on a boat. Much has changed following the terrorist attacks of 9/11, and although one still encounters the occasional foreign tourist in Gilgit-Baltistan (GB), they by and large comprise dedicated mountain climbers or hard core trekkers who are unfazed by any terrorist threats.

On the way to Khunjerab, the towering mountains seem to close in on the Hunza River with the road winding through some of the highest peaks in the world. The KKH steadily climbs up the Khunjerab Valley that begins from Sost and ends at Khunjerab Pass which at 15,500 ft is the highest border crossing in the world. Mir Salim Khan, the powerful ruler of Hunza State (1804–34), established complete authority over the area by extending the boundaries of his state across the frontier to China.[2] Khunjerab became his personal summer pasture thereby deriving the name 'Valley of the Khan'. Our trusty driver, who flew across the twisting KKH at a speed that went up to 80 km an hour, informed us that in local folklore it means 'Valley of Blood'—the combination of the two words *khoon* (blood) and *jerab* (valley)—as many travellers on their way to China have lost their lives in this treacherous region. I noticed that even he had to slow down the car a trifle in the Khunjerab area.

Almost an hour from Sost, we spotted a sign on a checkpost on the KKH manned by the Khunjerab Villagers Organization (KVO), the local NGO helping to guard the wildlife in the park. It informed us that we had arrived at the entrance to the Khunjerab National Park. Our driver got us registered (as all vehicles going up to the Khunjerab Pass have to be) and we entered the park's buffer zone. A few kilometres into the narrow gorge through which flows the Khunjerab River (which later joins the Hunza River), our car came to a grinding halt; a fast-flowing stream had cut across the KKH. We got out to gauge the threat. The ice-cold water was gushing down the mountainside carrying large boulders down to the river below. Not even a four-wheel drive vehicle could risk crossing it as, given the force of the water, a single slip would push it into the river below. In Karimabad at the Hunza Darbar Hotel where we had been staying, Salim Khan, the direct descendant of the original Salim Khan of Hunza State (whose family owns the hotel), had happily informed us that in the event of our ever deciding to take a dip in these glacier-fed rivers or nullahs, our prospects of survival were zero. 'Try putting your hand into these waters ... your fingers will freeze in 30 seconds or so,' he had exclaimed. He is a frequent visitor to Khunjerab; and although his family handed the territory to the Government of Pakistan long ago, they still own pastures and grasslands in Dhee nullah.

A melting glacier high up in the mountains was obviously causing the road to flood. Our driver suggested that we return to Sost and come back early in the morning when the water level would be lower (at night, the temperature drops and the glacier ceases to melt) to attempt a crossing. The thought of returning to Sost, a somewhat seedy little town located in probably the least picturesque spot on the KKH, was not an inviting prospect. I suggested we wait a while and, sure enough, as it began to get dark it appeared that the water was receding. Finally, inspired by the locals who nonchalantly hopped and jumped across the stream to the other side of the road, sometimes even losing their shoes in the process, I decided we too could pass the crossing on foot. We were booked at the only government rest house in the Khunjerab National Park so we had a place to stay overnight. We gathered our luggage, told our driver to come back in the morning, and gingerly made the crossing with the help of some of the local people. What immediately strikes you in the Gilgit-Baltistan region is that the people are extremely friendly, helpful, and gifted with a great sense of humour. In these rugged mountains, people have learnt to be cooperative and to laugh at adversity since it is the only way of ensuring their survival.

Our shoes and jeans soaking wet, we hitched a ride on the other side of the road and eventually arrived at the rest house at dusk. The rest house is adjacent to the next checkpost which is the official, manned government checkpost to the park. Khunjerab National Park was established in 1975 on the recommendation of the American biologist, Dr George B. Schaller, with the primary purpose of conserving endangered wild animals such as the Marco Polo sheep, blue sheep, the snow leopard, and other animals and natural resources in the region. As the entire area had been continually and traditionally used by local communities to graze domestic livestock, Dr Schaller recommended a 12 km portion of the park to be closed to grazing to protect Marco Polo sheep against intrusion and food competition. The selected section was declared to be the park's core zone.

Imposing a ban on grazing livestock, without buying the rights, created serious conflicts between the park management and the local people.[3] These issues were eventually resolved and signed agreements exchanged between the government and the local inhabitants. The local people now participate in helping to guard the wildlife of the park. Indeed, the Khunjerab Villagers Organization now provides guides to interested visitors to the park who want to visit neighbouring valleys, such as the remote Shimshal Valley, in the park's designated area. The park covers about 2,270 sq. km encompassing glaciers, alpine pastures, plateaus, streams, and ravines.[4]

A proper management plan for Khunjerab National Park was initiated in 1993 although it began being implemented in 1998. Today the park has its own directorate and staff in place, and is managed by the provincial wildlife department of Gilgit-Baltistan. The Chief Conservator Wildlife heads the park along with the support of three wildlife management officers. Currently, the management plan for the park is being revised by WWF-Pakistan; the new document is expected to be ready soon, having been reviewed by the relevant stakeholders. The park is considered to be one of the key biodiversity areas in the cold desert ecoregion of Pakistan. As many as 24 species of animals living here have been listed in the IUCN Red Data Book and CITES appendices as endangered, vulnerable, and low-risk species. Marco Polo sheep, blue sheep, Siberian ibex, snow leopard, brown bear, wolf, golden marmot, lynx, red fox, and Cape hare are the key species here.[5]

At the rest house, we met a young cyclist and environmental lawyer from the US, Jason Patlis by name, who was planning a four-day trip to a remote valley in the park in quest of the Marco Polo sheep which are now rarely sighted in the area. Unfortunately, as the KKH was completed in the 1970s, hunters have enjoyed easy access to the wildlife in the park. Marco Polo sheep in particular are highly valued for their meat and as trophies, and were almost hunted into extinction during 1960–70 when the KKH was under construction. Sadly the hunting of large males still continues in some areas.

Marco Polo sheep had been entering Pakistan from China via the mountain passes but now the sheep are confined to only a limited lambing habitat in the Khunjerab National Park. The KKH, with all its roaring traffic coupled with the border fence stretched partially across the valley along the Pakistan-China border, has greatly reduced the free movement of the sheep. Their numbers have recently increased on the Chinese side probably due to better protection by Chinese authorities. However, Jason was assured by the guides from the KVO that they could take him to a remote valley where several of these creatures had been spotted in recent weeks. I bade him farewell somewhat enviously since there was just no time in our hectic schedule for a four-day trek to see the Marco Polo sheep.

Fortunately, we did manage to spot some other elusive wildlife. At around five o'clock in the morning, we were awoken by the government wildlife guards. We crawled out of our VIP rooms (no water, no electricity but otherwise a sprawling three rooms suite) and went outside. The guards had set up a tele-spotter (high-powered binoculars); I managed to spot several Siberian ibex, a wild goat species, when peering through the lens. High up on the ridge of a nearby mountain, they were locking their magnificent horns (also valued as a trophy), and gambolling in the fresh cool air of the early morning. We watched in awe for several minutes, glad that the experience had proven to be well worth waking up for at 5 a.m. The ibex are also facing threats from overhunting since their trophy can sell for as much as US$3,000 for foreigners and Rs 60,000 for Pakistanis. WWF-Pakistan and IUCN have begun strictly controlled ibex trophy hunting projects in certain selected valleys of Gilgit-Baltistan. In these schemes, the ibex population is carefully documented and a mature animal is selected for its trophy. The hunter pays a special fee, 80 per cent of which goes to the community while the balance of 20 per cent goes to government. In Khunjerab, hunters can even obtain special permits to shoot ibex.

After breakfast, we departed for Khunjerab Pass, located almost an hour and a half away. Several European cyclists who had camped in the garden of the rest house were also up and they too left for the pass, albeit on their bikes. Prior to 9/11, many Americans and European cyclists used to come every summer to undertake this grueling ride up to the Khunjerab Pass. I don't know how they could do it with the air getting scarcer the higher the altitude they reached. Indeed, at the top of the pass, it is difficult to breathe and some people even pass out due to lack of oxygen. When I queried Jason, he replied: 'Very slowly ... we cycle up there very slowly!' We overtook quite a number of cyclists that morning, in fact, I was tempted to offer each a ride to the top.

The landscape gets bleaker as one climbs up the winding road to the Khunjerab Pass. There is not a soul to be seen for miles as there are no villages within the park's core zone. On the way up, we spotted dozens of golden-orange coloured marmots, cute furry rodents who did not seem to be bothered by the traffic on the KKH (which was sparse that morning), coming right up to the road. We also saw some more ibex grazing on a distant ridge. In winter, I was informed, the ibex come right down to the nullahs and can be seen clearly from the road; in summer, it is hot so they climb high up to the mountain ridges. We stopped frequently on the road, hoping to spot some brown bears since we were told that a few lived in the park. We were, however, out of luck; and it is virtually impossible to see the snow leopard which prefers to live on the snow line of the peaks.

It is estimated that only 200 snow leopards exist in Pakistan's northern mountains. In December 2012, a female snow leopard was found by local villagers in the Khunjerab National Park after being abandoned by her mother. Snow leopard cubs cannot survive in the wild once taken from their mothers because they teach the cubs how to hunt. The wildlife department of GB was alerted by the villagers and the cub was moved to a flimsy cage in Sost. Fortunately, the Snow Leopard Foundation in Pakistan raised funds to construct a purpose-built enclosure for her up in the picturesque Naltar Valley which was recently opened to the public.

On our way up to the Khunjerab Pass, we also saw plenty of yaks grazing next to the road. However, as they are domesticated animals, the sight did not unduly excite me. According to the latest study conducted in the park in 2011, which was sponsored by WWF-Pakistan, a total of 1,682 animals were sighted from seventy-six vantage points surveyed. Ibex was the predominant species with a total count of 899, followed by Marco Polo sheep at 716, and blue sheep at 67. Snow leopards cannot be surveyed easily and it is estimated that only a handful of them live in the park.[6]

Eventually, the road straightened out and we realized that we had reached the top of the mountain pass. Someone had described Khunjerab as egg-shaped and we were now right on the tip. Around us, several snow-covered peaks rose up majestically while the grassland of the pass itself was a verdant green, dotted with small ponds and covered with patches of purple coloured flowers that were just beginning to bloom. Apparently, in July, the pass is completely carpeted with flowers.

We arrived at the last Pakistani checkpost before the Chinese border itself and as the car came to a stop, I don't know whether it was from the lack of oxygen or some other such factor but I began to feel very drowsy. It was cold too, with strong gusts of wind blowing. We ran into a cyclist who was beaming triumphantly, turning his bike around from the cement pillar that proclaimed the border (there was a Pakistani inscription on one side and a Chinese one on the other at the time). I was tempted to ask him whether he had encountered any breathing problems but decided against tempering this joyful occasion. We took several photographs and then, in an attempt to escape the cold, tried to enter the Chinese built watchtower. However, it was locked so we had to turn back; Khunjerab Pass is truly spectacular but 15,500 ft is not exactly hospitable. We bundled ourselves into the car, regretting having to leave such a wild and awe-inspiring locale but glad that the high altitude and freezing temperatures ensure the inability of man to ever settle in this area. Some places deserve to be left to the wild.

Unfortunately, this wilderness will soon see lots of commercial activity as the China-Pakistan Economic Corridor (CPEC) will pass through this high border area. Already the first Chinese shipments are rolling in over the high mountains from Kashgar into Gilgit-Baltistan. One cannot help but be concerned about all the impact these trucks and lorries will have on this fragile mountain ecosystem and its precious wildlife.

Notes

1. Michael Palin, *Palin's Travels: Himalaya with Michael Palin* (United Kingdom: Phoenix Books, 2004). Website: <http://palinstravels.co.uk/book-3493>
2. *Hunza (princely state)*. Website: <https://en.wikipedia.org/wiki/Hunza(princely_state)>
3. UNDP-Pakistan, *Green Pioneers Stories from the Grassroots* (Karachi: City Press, 2002).
4. WWF-Pakistan, *Khunjerab National Park Management Plan* (2013).
5. *The IUCN Red List of Threatened Species* (2016). Website: <http://www.iucnredlist.org>
6. Rahmatullah Qureshi et al., 'First Report on the Biodiversity of Khunjerab National Park, Pakistan', *Pakistan Journal of Botany* (2011).

2

Shandur Pass: Endangered Wilderness

OUR TRIP FROM MASTUJ TOOK ALMOST THREE HOURS OF
tricky mountain driving in an open jeep to climb out of the last
green valley, which now lay far below us. We were in Chitral
district, on our way to Shandur Pass. I was excited as it was my
first visit to what is considered the highest polo ground in the
world, and that too just on time for the Shandur Polo Festival
which is held every July. We were travelling in an old Willy's
jeep which we were told is the sturdiest vehicle thus best suited
for these jagged Hindu Kush Mountains. The World War II
era jeep climbed steadily up to 12,200 ft and soon we levelled
out on to a grassy plateau with a large blue lake spread before
us encircled by rugged, snow-capped mountains. There are
actually three lakes on the plateau which are fed by melted
snow from the peaks that rise to a height of 2,000 ft above
the pass.[1]

We had arrived at Shandur Pass which lies between
the regions of Chitral in the Hindu Kush Mountains and
Gilgit in the Karakoram Mountains. The Chitral Scouts
have constructed a small three-room rest house, and a
permanent polo ground near the bigger lake called Shandur
Lake. Although the three-day festival had not yet begun, this
normally desolate site was already bustling with activity: stalls
were being set up, roadside cafes had their stoves smoking,
and gradually a trickle of vehicles was arriving, raising clouds

of dust on the plateau. As many as 15,000 people were expected for the polo tournament, and shopkeepers from the surrounding villages had already moved up to sell rugs, blankets, and food.

We were warned about altitude sickness and sure enough Ayesha Vellani, the photographer who was accompanying me, began complaining of migraine. We headed to the army camp located next to the Chitral Scouts rest house where a tent had been arranged for us. It actually proved to be quite spacious and comfortable with warm bedding and even a private tented toilet for us nearby. Women were nowhere to be seen at the polo tournament so both of us were quite a novelty and drew many curious glances. While Ayesha rested, I set out to look for Colonel Khushwaqt-ul-Mulk who happened to be Zainab Alam's (a friend of mine from Lahore) grandfather. The latter was the last prince of Mastuj, and a regular visitor to Shandur until his demise in 2010.

The Shandur Polo Festival takes place between teams from Chitral and Gilgit; during the three-day event, this isolated wilderness undergoes a dramatic change. Although it has been declared a protected area and is considered a national park, the effect of this annual tournament on Shandur's fragile natural environment has extracted a heavy price over the years. The government, however, is keen to promote the site as a tourist destination, and while we were there the then Prime Minister of Pakistan, Shaukat Aziz, was due to arrive as the chief guest on the final day of the tournament. By ignoring the environmental impact that is the result of the traffic the tournament attracts, the government and army officials are neglecting to safeguard the fragile ecosystem of the pass from human pollution.

Before the tournament began becoming a regular event in

the mid-1980s, Shandur Lake, which lies in a depression not more than 15 ft deep, was an important refuge for migratory birds as it lies along an international migratory bird route. For centuries it served as a high altitude wetland and a safe habitat for rare ducks like the garganey teal and the ruddy shellduck. Shandur Lake's very isolation and location on a high, flat plateau made it an ideal stopover for migratory birds.

With the establishment of a regular polo tournament and a permanent rest house on the pass (undoubtedly used by hunters), Shandur Lake could no longer offer a pristine haven for the ducks which eventually stopped nesting there. According to Ashiq Ahmed of WWF-Pakistan, 'Local hunters report that the ruddy shellduck, which was often spotted in Shandur, has not been seen in the Chitral area for the past two decades.' During the polo festival, which is usually held in the beginning of July, a few birds can be seen on the lake but these are mostly coots (waterfowl).

Polo is a passion for the local people and thousands come from all over the northern mountainous region to watch the matches every year, often trekking for days over the high mountains. They are allowed to put up their tents anywhere at a distance from the lake and polo ground. Polo has been played on the Shandur Pass for centuries though not on the commercial scale that one witnesses today. 'One hears of stories and poems written about polo in Shandur going back many years but the tournament itself began to be held periodically from 1925 onwards,' recalled Colonel Khushwaqt-ul-Mulk, the ninth son of Sir Shuja-ul-Mulk, the last ruler of Chitral. In 1913, the British gave Mastuj to Shuja-ul-Mulk who passed it on to Colonel Khushwaqt. At the time I met him, the Colonel was almost 90 years old but in excellent health; he lived in an old fort in Mastuj, an imposing structure with high mud walls and watchtowers. Today there are lovely wooden chalets built there in the orchard, under the towering maple and oak trees, where one can stay by booking a room at the Hindu Kush Heights Hotel in Chitral.

Colonel Khushwaqt was born on 13 June 1913. At the age of thirteen, he was sent to be educated at the Prince of Wales Royal Indian Military College at Dehradun. Once he had completed his schooling, he was commissioned in July 1935. The following year he was appointed to the Indian army and joined the 19th Hyderabad Regiment. In the meantime, following his father's death in 1936 and the accession of his eldest brother, Colonel Khuswaqt became governor of upper Chitral from its centre in Mastuj.[2] After retiring from military service and later from civilian employment, he spent most of his time at his home in the fort at Mastuj where he generously entertained visitors from all over the world. He also played polo in his youth and only gave up riding when he was well into his eighties.

Early the next day he asked us whether we had the time to visit his family's land further up Shandur Pass where they often went camping. We readily agreed and woke up to a beautiful blue sky after a blistering cold night. The temperature can suddenly drop at these heights so, from being burnt by the hot sun during the day (sunscreen is a must as one's skin literally peels off), one has to brave freezing temperatures at night when the strong winds howl through the pass. Colonel Khushwaqt approved of our Willy's jeep, saying he had one exactly like it, and that he didn't trust the new modern landcruisers. We set off to his picturesque meadows located past the practice ground, further away from Shandur Lake, where the polo teams were up early galloping their horses and practising shots. We had a lovely picnic on the grassy meadows dotted with wild flowers, and drank water from the bubbling streams that criss-crossed these gentle mountain slopes. He regaled us with romantic stories of moonlit polo matches played by the British at Shandur Pass.

After we returned to the camp, I set out to meet his son, the wiry Sikander-ul-Mulk, captain of the Chitral team and an ace polo player. He was camped out with his team members and horses at the other end of the sprawling tented village that seemed to have sprung up overnight. He told me:

All this activity has had a bad effect on Shandur Lake. The festival has not been properly managed. The public is allowed to come in and camp anywhere (although away from the lake)—they pollute the streams that feed the lake. Eventually, all the pollution ends up in the lake. Proper toilets with pit latrines should be built and trash cans for garbage should also be put up. Unfortunately, the organizers' main concern is for the handful of VIPs. No one cares about the 10,000 or so people who also come to see the match. We need to educate these people, make them aware of things like not throwing plastic bags into the lake, etc.

During the tournament, however, I noted that announcements were periodically made urging the spectators to respect the environment and keep Shandur clean but there was no one to enforce this and people could be seen littering the small streams that are found in abundance on the pass. It was on the last day of the tournament when all the activity came to a feverish crescendo, and helicopters flew over the pass bringing in the VIPs. They landed on a helipad near the lake where jeeps were waiting to convey the visitors up to the Chitral Scouts rest house. The sounds of drums beating echoed throughout the camp.

It was a hot day and we sat on comfortable chairs on the steps of the VIP area adjacent to the Chitral Scouts rest house which directly faces the polo ground. Colonel Khushwaqt had kindly invited us to watch the final with him. The Prime Minister had arrived earlier along with a couple of generals and their families in the helicopters. One of the wives who couldn't breathe properly in the rare atmosphere was being administered oxygen (it's far better to drive up to the pass after spending the night in Mastuj as that enables you to acclimatize to the high altitude). After having spent the last three days camping on Shandur Pass, my skin was burnt and I was in acute need of a hot shower. It was unanimously agreed that we would leave once the match was over.

Sikander-ul-Mulk was the captain of the Chitral team, while the Gilgit team was captained by the legendary Bulbul Jan; both men were in their early fifties yet in great physical shape. Bulbul Jan had been a very successful player in the freestyle polo that is played in this mountain region, having guided his team to victory for several years. It was a closely contested match that day between the 'A' teams of Chitral and Gilgit, and I must confess that I was not greatly enamoured of the 50-minute long, no holds barred, sport. Freestyle polo is quite rough on the players who are hitting and obstructing one another; sometimes the horses even collapse and die suddenly at this altitude. A crowd of almost 15,000 people had gathered at the ground to watch the sport, and during the long interval they cheered on the ceremonial dancing and drumming. It was a close second half but eventually Gilgit won the match.

After the tournament concluded, I took a quick walk around the lake and found plastic water bottles and juice boxes, tin cans, and plastic bags littering the edges of Shandur Lake. From up close, the water of Shandur Lake appeared to be quite muddy and teeming with tadpoles, and I realized that the blue colour the lake assumes from a distance is merely the reflection of the sky. Researcher David Johnson, who completed his MSc dissertation on Shandur Lake while studying at Oxford University, has written: 'The festival directly affects the primary values of the wetlands by deteriorating the aesthetic attributes of the lakes and impairing the provision of drinking water and pastureland.'[3]

He listed a number of management actions the government can take to safeguard Shandur from further desecration. These include a system for the collection and removal of refuse, the establishment of buffer zones, and provision of proper toilets and washrooms. He also recommended that the local people be involved in the management of the lake's environs. The Pakistan Wetlands Programme provided David Johnson with technical support to conduct this research. He spent several days at Shandur sampling the water in the lakes; testing its chemical and physical properties before, during, and after the polo festival in 2006 in order to assess the impact of the festival on the lakes.

Closer to the lake, I could see the streams which feed Shandur Lake, and some came from the direction of the tented city. There were also many dead carcasses (of yaks and goats) lying near the lake but I was unaware of the cause—whether they had died from some disease or from drinking the polluted water of the lake. The local people use Shandur as a summer pasture for their precious livestock and live up on the mountain slopes around the lake in stone houses which protect them from the strong winds.

Earlier, we had gone up to visit one of these summer

settlements and found them to be by and large populated
by old women and young children from the nearby villages
who had brought up their livestock for summer grazing. The
stone settlement built up high on the ridges seemed to be a
village lost in time; the only signs of modernity were the plastic
bottles being used by the inhabitants to store water.

The people camping out in Shandur were, however, aware
of the lake's pollution and weren't using water from the lake
for drinking purposes. Indeed, there are a great many clean,
glacier-fed streams coming down the slopes of the pass that
act as a better substitute. Some of the people who had been
coming regularly for years to the polo tournament stated
they no longer bathed in the lake because the water causes
irritation of the skin.

Before sunset, the large tented city began to rapidly
disappear in a whirlwind of dust. A long line of jeeps and
people could be seen making their way out of Shandur in
both directions: towards Chitral and in the opposite direction
towards Gilgit. Indeed, people began departing straightaway
after the prize distribution ceremony was over, leaving all
their garbage and discarded items behind. The wind suddenly
began howling and all the dust rose into the air; I couldn't
believe I was witnessing a dust storm at 12,000 ft! I rushed
back to our tent to gather our luggage and have it loaded on
to the jeep so that we too could immediately leave.

I ran into Sikander-ul-Mulk, who looked a bit crestfallen
after losing the final, and he told me that he was waiting for
darkness to fall before he left with his teammates; the losing
team usually leaves much later to avoid the disappointed
crowds. I pointed out the brewing storm which would
undoubtedly result in freezing temperatures at nightfall and
his mournful reply was: 'See, even the spirits who live here are

so upset at Chitral losing that they are blowing up a storm!' Sans Sikander-ul-Mulk, we jumped into our jeep, grateful for the means of exit, and joined the long line of vehicles heading down towards Mastuj. Stuck in traffic jams, it took us several hours to reach Chitral town. By nightfall, Shandur would once again become a silent, desolate place.

I've been back to this soiled paradise on a more recent road journey over the mountains from Chitral to Gilgit in late summer. Shandur Pass looked much cleaner, although we didn't linger too long because we were keen to reach Gilgit town before dark. I was told that the Shandur Polo Festival in recent years has fallen prey to security concerns, floods, and ownership disputes. At least thrice since 2010, the Gilgit team has not participated in the festival to protest against the federal government's decision to assign the management of the event to the Khyber Pakhtunkhwa province (where Chitral is located). There is also a dispute regarding the clarification of Shandur ownership rights and the government they belong to. In 2015, when the government of Gilgit-Baltistan was eventually persuaded to end their boycott, the Shandur Festival was again cancelled due to massive flooding in Chitral district. The Gilgit team has been sorely missed at the tournament, however, and the government is hoping to revive Shandur's potential to attract large numbers of local and foreign tourists in the years ahead.

However, some environmentalists like Ali Ahmad Jan, who hails from Gilgit-Baltistan, feel that the festival should be suspended for now, pointing out that it is an environmental menace that has become a bone of contention between the two administrative regions. He says that the cost of organizing the festival entails wastage of much public resources and, given all the floods and devastation that have hit Chitral district in recent

years, he would prefer to see these resources being devoted
to rebuilding roads, schools, homes, and other infrastructure.
Certainly Chitral, and even parts of Gilgit-Baltistan, are
currently facing the brunt of climate change and the government
needs to do more to help the people of this region get back on
their feet again.

Notes

1. WWF-Pakistan, *Coastal Lagoons to Alpine Lakes: A Guidebook to Pakistan's Significant Wetlands* (2012).
2. 'Obituary of Colonel Khushwaqt-ul-Mulk', *Telegraph* (Mar. 2010).
3. David Johnson, 'Cultural Eutrophication of Shandur Lake', MSc dissertation (Oxford University, 2006).

3

Kalash Tribe of Chitral: Struggle for Survival

THE KALASHA ARE THE LAST POLYTHEISTIC PAGAN GROUP in this Islamic region and are under pressure to convert to Islam. The scenic beauty of their valleys and the tribe's ancient culture is a magnet for tourism into the area. I had been to the Kalash Valley of Bumburet a couple of times earlier; however, this time I was going after massive floods had devastated the mountain region of Chitral in the summer of 2015. 'You won't recognize Bumburet Valley when we get there,' my driver informed me as we navigated through mountain streams and attempted to cross the broken roads that had been washed away by the floods. One now needs a four-wheel drive vehicle to reach the Kalash valleys, a harrowing two-hour drive from Chitral town in the Hindu Kush Mountains bordering Afghanistan.

The three Kalash valleys of Bumburet, Rumbur, and Birir are encircled by high mountains, a natural barrier from the Afghan province of Nuristan whose people were converted to Islam in the nineteenth century by the Afghan King Abdur Rahman. The Kalash are the last survivors of Kafiristan (of which Nuristan was once a part), and their pre-Islamic religion is focused on their environment; although they worship a creator called Dezau, they also believe that trees, stones, and streams all have souls.[1]

A narrow jeep track used to lead into Bumburet, the largest of the three valleys. Now it is completely inaccessible, strewn with large boulders brought down by the floodwaters. I had to hang on to my seat as our jeep cut through agricultural fields and criss-crossed the main nullah just to reach the villages beyond.

During my last visit, I had stayed at the Pakistan Tourism Development Corporation Motel, which was now in ruins, situated at the entrance of the Bumburet Valley. It was a well-designed building with a large garden, surrounded by a gushing water channel. Inside, the rooms were constructed of wood and glass. Clean and efficiently run, it had been quite empty in 2006. Due to the 'war against terror', this region had witnessed a drop in foreign tourist arrivals. When the summer floods struck in 2015, the gushing stream turned into a surging river with floodwaters rushing down the mountain sides into Bumburet Valley, destroying the motel's wooden chalets and large garden.

The rest of the valley's main town was also badly hit. Once a popular tourist destination, the big hotels had all suffered major damage, as had countless shops, around two dozen houses, many orchards, fields, and of course the main road itself. I remembered the main road from my previous visit. Although it had even then been a narrow jeep track surrounded on both sides by hotels, it had been packed with traffic moving up and down the valley. Much to my amusement, I had noticed several small Suzuki cars filled with bearded men from the Khyber Pakhtunkhwa province zipping up and down the road. 'They come to Bumburet to stare at the women,' one of the Kalash men had told me disdainfully.

In contrast to the other more conservative valleys of Chitral, the Kalash women go about with their faces uncovered and enjoy much greater freedom. 'We are free here—we can choose our own partners for marriage and we can even leave them if we want,' a Kalash woman had told me during my stay. How wonderful to be able to enjoy such freedom, not accorded to many women, in these mountain valleys that are surrounded by a conservative patriarchal society.

We headed towards the Kalash Cultural Centre built by the Greek government in 2004, which was thankfully still intact. Akram Hussain, a Kalash man in charge of the centre, explained:

> I would say the floods were the result of all the deforestation that has taken place in these valleys in recent years. Just look at the bare mountain sides; and of course climate change was also responsible. It is almost the end of September now and it is still warm; in earlier years the summer would end in August when we would celebrate our autumn festival.

He showed us around the spacious building constructed in their traditional style with stone and wood. The centre houses an impressive museum, a community area, and educational facilities for the Kalash people. The museum displays a large collection of Kalash antiques such as old jewellery, cooking utensils, sculptures of horses, rugs, and costumes. The Kalash women still wear their traditional attire: an embroidered black frock tied at the waist by a belt or sash. Until recently, the women's dress was woven at home and the yarn used to be dyed by hand. Now they buy their materials from the bazaar in Chitral and embroider them by machine. The men no longer wear traditional outfits, opting for the ubiquitous shalwar kameez. The women also wear innumerable beaded necklaces and the distinctive embroidered headdress.

Sturdy stone walls built around the centre by Greek volunteers ultimately saved it from the forceful impact of the floods. 'The high walls withstood the floodwaters, which came rushing down from the mountains at one end of the valley and swirled around the building,' said Hussain. Sadly, it is likely to be the last sizeable aid package that the Kalasha will receive from the Greek government, which is now itself facing a severe economic crisis at home. However, they continue to pay for the teachers' salaries and for the school supplies for around a hundred children studying at the Kalasha Cultural Centre.

The Greeks believe that the Kalasha are descendants of Alexander the Great's army which marched through the Hindu Kush Mountains centuries ago. For many years this was just a theory until, in 2014, a team of geneticists led by Oxford University sampled genomes from around the world and found that the DNA of the Kalash people of Pakistan closely matches that of an ancient European population. The Kalash are, therefore, in all likelihood the descendants of the ancient Greek-Macedonian armies which set up outposts in this region 2,300 years ago.[2]

According to a census conducted in July 2014, only 4,114 Kalasha survive today but there have been many conversions to Islam as an increasing number of outsiders settle in the Kalash valleys. According to Hussain, around 1,800 Kalasha live in Bumburet, but higher up along the sides of the valley. Their neighbours across the high mountains are the Taliban who now control large swathes of Nuristan in Afghanistan. When I was last in the Kalash valleys, a member of a local NGO had jokingly asked me whether I had come 'looking for Osama bin Laden' because at the time many had believed that he was hiding in the inaccessible mountainous terrain of Nuristan.

Although the Taliban have in the past threatened to convert the Kalash by force, the Pakistan Army is now patrolling the border. In fact, army checkpoints have sprung up in the Kalash valleys where ID cards are meticulously checked and several army jeeps are visible on the broken roads. Despite the measures taken, in the summer of 2016, attackers crossed over from the Afghan side and opened fire on a band of four Kalash shepherds. In this predawn attack in a remote pasture high up in the mountains, two shepherds lost their lives. There was an immediate outcry and the army moved in and cordoned off the area. They then launched a targeted operation in the region to hunt the militants, killing five of them. The border with Afghanistan has now been sealed off by the army.

According to Hussain, the Kalash face a multitude of problems:

> Our culture was already under threat, and now these floods have destroyed our crops and orchards. We will have to buy food from the bazaar and store it if we are to survive this winter. Luckily, there was no loss of life in Bumburet because we got a call from the border police which jointly patrols the border with the army. They called us over our mobile phones to warn us that the flood was coming, I called people living near the Bumburet nullah [the mountain stream that runs across the valley floor] and told them to get out ... We have never seen floods like this in the Kalash valleys before. I would say that at least half of Bumburet Valley was destroyed or damaged by these floods.

Notwithstanding the widespread destruction, only a few of the houses damaged by the floods belonged to Kalash families. 'Our houses are mostly built higher up and all the people living below ran up to our homes,' explained Shaheen Gul, a young Kalash resident of the nearby Krakal village, as she eagerly showed me around the place. 'But our fields with corn and beans that were ready for harvest, and fruit trees like walnuts and apricots, are gone as they were near the nullah.' Built high above the main road, Krakal village is one of the oldest settlements in Bumburet. Gul elaborated on the incident, stating that it had been a terrifying night for all involved. 'We could hear the flood before it arrived—we were so scared by the roaring sound in the middle of the night. Then the earth started shaking as if there was an earthquake. It was raining very hard that night,' she recalled. 'Later, when we came down, we saw all the destruction.'

The Kalasha homes are extremely sturdy and well built, and although the wooden houses are almost stacked upon one another, their unique style of construction with stone, wood, and mud ensures that they are earthquake-proof and, evidently, also flood-proof. Indeed, the Kalash villagers told me that when an earthquake does strike, they run into their homes to protect themselves from cascading boulders. The Kalasha are generally an open, friendly, and happy-go-lucky people who usually welcome outsiders to their valleys: their windows and doors always remain open in the summers and there is little crime here. The younger girls confided to me:

> It is difficult to live here during the winters when the valley is completely cut off from the outside world so no one can leave or visit. Also, there are no jobs here for young people but we do have schools now. Still, we wouldn't want to live anywhere else. We are happy here.

Across the road, outside Krakal village, is the *bashali* or menstruation house to which women are sent when they are considered to be in an 'impure' state. This also includes the period of childbirth. Purity is an important part of all religious rites and ceremonies in the Kalasha culture; the basic concept is that the deities will not receive prayers and offerings if a state of purity is not maintained. Other than their faith in Dezau, the village also believes in the goddess Jeshtak. There is an old temple there devoted to this goddess who is regarded as the guardian of all family matters. The Greek government had built the *bashali* in Krakal village; it is a clean, well built structure with spacious rooms and bathrooms built in the traditional stone and wood style. Fortunately, it too escaped damage from the flooding.

Climate change experts have blamed the erratic monsoon weather in 2015 on El Niño, a periodic warming of the ocean along the equatorial Pacific that impacts weather patterns. Some monsoon currents were able to penetrate the Hindu Kush Mountains, an area which generally lies outside the monsoon belt. The persistent rainfall accelerated snowmelt and triggered both floods and the formation of glacial lakes which devastated the Chitral district.

Rumbur Valley was the second area to be badly affected by the flooding; however, the third Kalash Valley of Birir was undamaged by the floodwaters. It took us over an hour to reach Rumbur from Bumburet by crossing even more broken roads. Mohammed Iqbal, a Kalash social worker with the Pakistan Red Crescent Society, was there to greet me when I finally arrived in Rumbur, shaken by the difficult journey especially the last section when our jeep had to cross a road on a cliff that was barely even a track. 'We had rainfall almost every day. There were two types of floods: one was a glacier flood in Rumbur nullah and the other was a flash flood gushing down one of the mountainsides,' he told me as we crossed a bridge that had been spared by the floods and walked into the main bazaar.

An early mobile alert from shepherds high up in the mountains prevented any loss of life in the valley. 'The shepherds told us that they saw cloudbursts over the Gangalurat glacier on the boundary with Nuristan which then burst,' he said, pointing to the high mountains behind the valley. The floods brought down boulders, trees, and everything else in their path, destroying orchards, fields, water channels, and roads.

'The sound of the floods was deafening. The children are still so traumatized. Now, each time it rains, they start crying and saying "a flood is coming",' said Naseem, a Kalash woman who lives in Grum village further up in the Rumbur Valley. She added that with the advent of winter, they would soon be facing some hard times. 'All our crops of potatoes, beans, and sorghum near the nullah are gone along with so many fruit trees,' she said. 'The drinking water supply has not been restored and the women have to go down to the nullah daily to fetch water and carry it up in large vessels. The electricity has not been restored either.' A micro-hydel plant used to provide electricity to Rumbur Valley but was now rendered useless due to the damage caused by the floods. The community, however, did plan to fix it before winter. By now the sun was beginning to go down and, since there were no hotels in Rumbur, I had to leave for Chitral town in the jeep.

I was staying at the comfortable Hindu Kush Heights Hotel in Chitral town where I had met Quaid-e-Azam, a Kalash man from Rumbur who works for the hotel. He told me that despite the obvious hardship, the Kalash people prove to be quite resilient; even celebrating their annual Uchaw Festival in late August. 'Uchaw commemorates the harvest,' explained Azam, 'and we celebrated it despite all the destruction. This is a religious festival and we have to continue with our culture and rituals.'

A Kalash man who works for the government's Tourism Corporation in Chitral town, Zarin Khan, later told me:

> During the Uchaw Festival, our elders prayed to Dezau to save us from future floods as the people were very worried. There has been so much damage caused by the flooding. The Kalash people also feel that the deities are angry with us because, with all this deforestation in our valleys where trees and entire forests have been cut down for purposes both of construction and firewood, the animals living up in the mountains have no pine nuts to eat and even the birds go hungry. We feel the floods are coming as they are cursing us people for all this deforestation and destruction of their habitat.

Azam was deeply worried about their future. Sitting behind the reception desk of the hotel, he said in a pensive note:

> The floods are becoming worse each year. Today, because of the growing population in the Kalash valleys, people have no choice but to build near the nullah but they are risking their lives. They were only saved this time because of the mobile phone warning. They were extremely lucky ... No one is safe now with all this climate change. The next few years are going to be very hard for us. We can only hope to survive with better planning.

Chitral has seen even more destruction since my last visit. More flash flooding hit the district in the summer of 2016 killing over thirty people although the Kalash valleys were spared this time around. In 2017, over two dozen people were killed when avalanches hit remote parts of the district in the late winter. Chitral remains extremely vulnerable to different types of disasters thus a localized system of disaster management needs to be developed on an urgent basis in order to ensure minimal damage in the years to come.

Notes

1. Luke Rehmat, 'Do you know the Kalasha tribe of Pakistan?', *The Kalasha Times* (Jan. 2013).
2. Nicholas Wade, 'Tracing Ancestry, Researchers Produce a Genetic Atlas of Human Mixing Events', *The New York Times* (Feb. 2014).

4

Palas Valley: Wealth of Biodiversity

PILES UPON PILES OF TIMBER LINED THE KARAKORAM Highway (KKH) as we drove up to Pattan, located on the banks of the Indus River. An hour's drive from Besham, it is here that travellers who are on their way to Gilgit often stop. It was the year 2007, and we were greeted by a pile of logs every time we turned a bend in the road. This continued for quite a while as we drove in silence, paying our last respects to the thousands of trees that had fallen victim to the carnage. It seemed as if an entire forest must have been felled to obtain such a quantity of tree trunks; the piles were almost 10 ft high and appeared to be deodar wood. A horrifying silence encompassed the car upon the realization that a deodar tree takes almost a hundred years to mature.

'Yes, there must be over 100,000 trees piled up on the road. They claim they cut these trees way back in the 1990s before the ban on logging was enforced in the NWFP [today known as Khyber Pakhtunkhwa province], but it is certain that there are many newly cut ones too,' informed the members of the NGO, Palas Conservation and Development Project (PCDP). We were visiting one of their projects based in Palas *tehsil* in Kohistan district. Palas is home to Pakistan's most important remaining tract of natural West Himalayan forest, hence it is recognized as a global priority for the conservation of biodiversity.

31

Only a year or two earlier, the timber had been hauled out from Palas Valley and nearby forested areas then lined up on the road. It was gradually being sold to contractors who would pay Rs 60 per tree whereas, in Lahore, the same quantity of wood was worth Rs 2,200. The Kohistanis were being exploited by the contractors but has recently begun to realize the real value of the precious forests which are spread over their mountainous ridges. 'The people of Palas are becoming aware now. They've realized their error in selling their forests. Now some villagers are so wary of outsiders that they won't even let the Forest Department into their community-managed conifer forests,' said Rab Nawaz, a wildlife expert who was at the time seconded to the PCDP from WWF-Pakistan. He had worked in Palas Valley for over a decade and earlier with the World Pheasant Association (WPA). With his blond hair and blue eyes, I initially mistook him for a Pathan only to later discover that he was actually an ecologist from Wales who had moved to Pakistan to work for the WPA in 1994.

Before Rab Nawaz converted to Islam in 1998, he was known as Rob Whale. He worked with the WPA in Abbottabad, breeding captive pheasants, before moving to Palas Valley to observe and record changes in pheasant numbers in the wild. The Himalayan pheasants are among the most beautiful birds in the world but are very shy by nature. In Pakistan, the birds are largely to be found in the remote north of the country including Palas Valley which alone boasts of 5 different species of pheasants: the koklas, monal, western tragopan, khalij, and the chir peasant. Of these 5 species, the western tragopan, which can be found in Palas, is perhaps the most magnificent and the rarest. The western tragopan was first filmed by a documentary team led by Rab Nawaz under the WPA's Pakistan Galliforme Project.[1]

Rab Nawaz completed his integration into Pakistan by marrying a Pakistani woman in 1999. Today they live with their children in Islamabad where he now works for WWF-Pakistan. According to him, putting down these roots has helped strengthen his links to his adopted country. Palas Valley, in particular, has inspired him with its beauty. 'Palas is very special to me,' he said. 'Given its inaccessibility, its forests and rivers are undisturbed.'

The WPA had been working with other partners to support a wide variety of conservation and development work in Palas Valley. The aim was twofold: to conserve the forests and the rich diversity they encompass and also to help the people living there. The organization supported several projects in the area and also provided aid after a devastating earthquake that convulsed the region in 2005. It also assisted the PCDP which was implemented from 1998–2007. Earlier, the WPA had supported a project called the Himalayan Jungle Project, implemented from 1991–95, which paved the way for the PCDP. The Himalayan Jungle Project had concentrated on the western tragopan population in upper Palas. The western tragopan was sighted and then captured on film in the valley by the wildlife cameraman, Reza Abbas, who has since passed away in Karachi. Surveys by the WPA have suggested that these forests contain an estimated 300 pairs and are considered to be among the largest populations of western tragopan in the world.

The PCDP worked hard to raise awareness about the value of the forests in Kohistan and safeguard its wildlife and plants by enabling the local community to tackle both poverty and natural resource degradation. This belated cognizance of the importance of their environment helped convince the 30,000 inhabitants of Palas to protect their forests from outsiders and enable new trees to grow by controlling grazing. 'If we can save our forests, it is better for us in the long run,' said Ghulamullah, a local villager. 'Slowly, change is coming to Kohistan.'

The Kohistanis are, of course, fiercely opposed to outsiders entering their valleys and had been resistant to change for many years. Initially, Rab Nawaz had a hard time winning over their trust. One cannot simply walk into Palas Valley unless well acquainted with the local villagers. The local people are by and large of a suspicious bent: their homes are perched on hilltops and yet even these have small watchtowers with gun turrets. Feuds over landholdings and women are commonplace and can carry on for generations. Almost every grown male carries a gun. I recall asking one of the local men, 'What is the necessity of all this feuding? Your lives are already so tough.' His haunting reply was, 'We are human, not animals; we only pick up our guns and use them when we must.'

Women here have few rights and observe strict purdah. They are not permitted to mix with the men, although they work equally hard in the fields and have to walk for miles to fetch firewood and spring water. There are no schools for girls and only a few primary schools for boys. The people are staunchly religious and follow an austere interpretation of Islam. In short, this is a highly conservative and closed society. However, the twenty-first century is beginning to intrude: just across the bridge over the Indus River lies the KKH which has made accessible many other remote mountain valleys in this region. Even so, change can sometimes be dangerous. In 2012, a video filmed on a mobile phone in Kohistan began circulating on the Internet showing four women and a teenage girl 'dancing' with men at a wedding. As a result of partaking in this activity with male partners not related to them, the females were killed. The video depicted well-clad women, with only their faces visible, clapping and singing for two brothers who were dancing at a wedding ceremony. Deeming it scandalous behaviour, the tribal elders ordered them to be murdered to protect the tribe's 'honour' whilst the two brothers escaped from the area and went into hiding for fear of their lives. These 'honour' killings in Kohistan shook the nation, not only because of the senseless murders but also because of the mystery surrounding the case and the cover-ups that ensued. To date, the perpetrators have not been brought to justice.[2]

Palas Valley, its high mountain ridges shrouded in cloud, is certainly a remote and mysterious place. It has long been known in scientific circles, both in Pakistan and abroad, for the richness of its biodiversity. Half the valley, i.e. 1,300 sq. km, lies within the monsoon belt and the other half outside it. The range in altitude implies a great variety of habitats, from subtropical to alpine. Scientific surveys conducted between 1987 and 1995 in Palas have identified over 500 varieties of plants growing in the valley. According to Rab Nawaz, it is also home to some 162 species of birds and 33 wild animals.

The inhabitants of Palas have always respected their natural environment and consider it unlucky to cut a green tree in spring. However, their need to feed their children comes in the way of their belief. Kohistan is a poverty-stricken region where jobs are rare and every day is a struggle for survival. People own small plots of terraced land and live off what little they are able to grow. Many men migrate to Karachi and other large cities in search of work. Despite all the troubles, the mountain people have managed to survive for centuries in this harsh albeit beautiful environment.

It was during our visit to Sherakot village in Palas Valley when I met a bright, young schoolboy who had won an award for his haunting poem in which he compares how life is expensive but death is cheap in Kohistan. In 2007, the WPA built and funded a school in Palas to provide education and encourage the participation of the entire community in the conservation of the valley. I met the schoolboy at this school, an impressive structure that was well-equipped with books and wildlife posters.

I also met some of the women who had gathered in the private rooms of one of the houses I visited. These rooms were in the lower section of the wooden house where the livestock was also kept and the secluded rooms faced an orchard and the fields beyond. The women were thrilled to see me, a female stranger in their land, and admired my clothes and shawl, asking me dozens of questions about where I was from. They were extremely friendly and hospitable, cracking open fresh walnuts and fetching me apples picked from their orchard. They reluctantly bade goodbye when I had to leave. To this day I wonder whether any of those women I had met that memorable day were victims of the 2012 incident and killed by their relatives for merely clapping their hands and singing in front of other men.

During my visit to the valley, I also witnessed the demolished wooden and stone built homes that the earthquake that struck Pakistan on 8 October 2005 had left behind in its wake. Although there was relatively little loss of lives as most of the people were out in the fields, almost 1,000 houses were destroyed. The houses near the forested areas were, however, saved as the trees had protected them from landslides. The people of Kohistan decided to go ahead and rebuild their homes instead of waiting for handouts from the government. The WPA and its members raised almost GB£20,000 for the people of Palas Valley to rebuild their homes. Inevitably, the reconstruction resulted in more trees in the area being cut for housebuilding purposes. We climbed up to a mountain ridge above the earthquake-hit village of Sherakot and I was distressed to see that more conifer trees had been freshly cut.

After half an hour's walk into the forest at a higher altitude, we arrived at a densely forested mountainside where the only sounds I could hear were the cries of birds. I walked further into the thick canopy and came upon a crystal clear mountain spring bubbling over moss-covered rocks. Sitting on a large rock, I drank from the clear water and inhaled the fresh mountain air while listening to the music of the forest. This is what the Japanese mean when they speak of a 'forest bath'; one's spirit feels cleansed and rejuvenated when connecting with nature in such a pristine environment. I was mesmerized and wanted to go deeper into the forest but since it was late afternoon, we had to head back down.

Unfortunately, we could not make it into upper Palas with its untouched forests replete with pheasants because the jeep track leading there had been damaged by landslides. The dense forests in this part of Palas have not yet been threatened because they are an entire day's trek on foot from the jeep track. Thus the timber mafia has been unable to enter the higher reaches of the valley.

We spent our two days in Palas visiting the nearby villages and learning about the Palas Non Timber Forest Products Management Project, which was also supported by the WPA. The idea was to provide alternatives to commercial timber felling. The project helped the local people grow fruit trees such as walnut and apple, enabling them to make some extra money. It also imparted knowledge on the various medicinal plants found in the forests which can be used for the treatment of both humans and livestock, and therefore have cash value.

The project provided some homes with honeybee boxes housing indigenous bees. The honey collected and sold in the local market helped contribute to the household income. Keeping in mind that the upper Palas was populated with many Chilgoza forests, the NGO also taught them how to extract the chilgoza nuts without damaging the trees, and then trained them to market and sell the nuts in the larger markets. 'Slowly, people are becoming aware of the benefits of all this training,' said Mohd Nawaz, a local villager. The process was slow but word was spreading throughout the valley. The local villagers understood that they could make more money by implementing on what they learned from these annual activities than by selling their timber every twenty years.

The PCDP had earlier established community-based organizations in many villages; training activities had been conducted on animal rearing, health, nutrition, and sanitation. Various elements of the infrastructure were also rehabilitated such as a suspension bridge, watermills, and irrigation channels. Considering Palas' largely traditional subsistence economy with a very low development index, these were necessary actions. Kohistan is one of Pakistan's least developed districts with a mere 4.8 per cent of its land area under cultivation.

However, it is due to its remoteness and lack of development that its forests and rivers have thus far been largely safeguarded. The forests of Kohistan can continue to be conserved in the future if the people realize the value of the treasure they contain within their thick foliage. Currently, there is some good news for Palas as the government's Pakistan Science Foundation has prepared a dossier to include the valley in UNESCO's Man and the Biosphere Programme. An application to this effect has been submitted to the Khyber Pakhtunkhwa government for approval; once approved it will be sent to UNESCO. A biosphere reserve is an internationally recognized conservation reserve created to protect the biological and cultural diversity of a region while promoting sustainable development. If Palas Valley is included in the programme, its future as a haven of biodiversity could be assured.[3]

Notes

1. 'Palas Valley—North Pakistan', *World Pheasant Association*. Website: <https://www.pheasant.org.uk/palasvalley>
2. 'Kohistan "honour" killings: Four years on no justice in sight', *The Express Tribune* (Jan. 2016).
3. 'Activities of MAB Programme in Pakistan', *UNESCO*. Website: <http://www.unesco.org/new/fileadmin/MULTIMEDIA/HQ/SC/pdf/MAB_national_report_Pakistan_MABICC28_en.pdf>

5

Ayubia Park: Home to Precious Natural Forests

AFTER A BREAK OF SEVERAL YEARS, I FOUND MYSELF AGAIN on the winding road to Nathia Gali in the summer of 2008. The hill station built by the British during their rule over the subcontinent is located in the lower Himalaya in Khyber Pakhtunkhwa province. Happily, the road was far smoother than when I had last visited, and the monkeys were found to be still sitting on the road when we passed through the stretch of the path that cuts through Ayubia National Park. The smell of pine was in the air and the forest looked dense. Indeed, I was relieved to see that nothing had changed dramatically since my last visit. After all, Ayubia National Park with its 3,312 ha of undisturbed, natural forests is a well-protected national treasure. It was a pleasant sight that more or less reflected my childhood memories of it, unlike Murree which is now a horror that I avoid visiting at any cost. Seeing its ill-planned mushroom growth of hideous concrete structures just breaks my heart. All the trees, parks, and lovely old cottages of Murree are now just a memory.

Pakistan's moist temperate forests (such as those in Ayubia) function like large water tanks, collecting water from the rain that falls in abundance in this high mountain region. The water is then released gradually in the form of natural springs. Ayubia's forests thus play the vitally significant role of watershed for Mangla and Khanpur dams. The water emanating from these forests also serves as Murree's water supply, and is also piped directly into the water tanks in Kashmir Point. In fact, there is the famous pipeline walk which one can take through the national park: a 45-minute long, level stroll that takes you deep into the forest. I have undertaken the walk many times since a trip to Ayubia was an essential part of spending my childhood summers in Murree. At 2,499 m above sea level, you meander along a trail on a mountainside through a thick coniferous forest. It is easy to undertake the 4 km walk and you can listen to the sounds of birdsong while gazing at the verdant valley on one side. It was in 1930 that the British built a water pipeline, starting from Dunga Gali where a large water tank was built, to carry the stored water to Ayubia and then to Murree. Since the pipeline was covered with earth and stones, the track began being referred to as the pipeline walk.

The following morning we headed to the information centre located within Ayubia National Park to learn about the problems the park was facing. This was a relatively new building with large windows and a terrace that belonged to the Wildlife Department, and was supported by WWF-Pakistan. Inside were several stuffed animals, such as monkeys and deer, and posters of the park. Representatives from Sungi, the NGO which was then working with the local communities to help conserve the forest, were there to welcome us.

First, we heard from Mohd Safdar Shah, then the deputy conservator of Wildlife NWFP, who told us about the park which was established in 1984 with around 1,600 ha of reserved forest. In 1997, another 1,600 ha were added. The park is now known for its outstanding flora, fauna, and scenic landscape; the natural forests are considered to be its unique feature. Although the park is government property, it is accessible to the public and has several marked trails that are open during the day.

Felling or damaging trees, hunting animals, and clearing or mining land is strictly forbidden inside the park. No house construction is permitted either nor do the local people have any rights over the park's resources. The park consists of subtropical pine species, moist temperate forests, and subalpine meadows with elevations ranging from 1,050 m to 3,027 m, the latter being the summit of Miranjani peak. It is estimated that there are around 80 medicinal plants in the area which have an enormous research potential. There are also 150 species of birds residing in the park, including koklas and kalij pheasants. Two species of flying squirrels[1] can be found here as well.

Unfortunately, creatures that were once found in abundance in the area: the monal pheasant, musk deer, and the black bear have been driven to extinction from the area. The principal threat to the park is from the local communities who resent being deprived of their traditional forest rights, including that of foraging for fuelwood as not everyone can afford LPG cylinders. They cut the grass within the park and bring their livestock to graze thereby allowing their animals to devour the young saplings as well. The thousands of tourists who visit the national park each year are also a hazard as their cigarette butts have ignited many forest fires. Indeed, whilst there we witnessed the guards and staff extinguish a fire started by some young men on the pipeline walk.

I have to admit that I have never got around to climbing Miranjani peak in Ayubia. However, one summer not very long ago, I managed to trek up to Mushkpuri top which at

2,800 m or 9,452 ft is the second highest peak in the national park area. The trek takes about three hours, throughout which you can enjoy the magnificent scenery. On the way to the top, you clamber through grassy meadows carpeted with wild flowers and lush green forests whose floors are matted with pine needles. At the flat top there is undulating green grassland dotted with thousands of daisies waiting to greet you. There are hundreds of butterflies too, and we even ran into a professor who was collecting them. We enjoyed our picnic admiring the bird's-eye views of Ayubia National Park and nearby Kashmir. I wondered why I had not undertaken this trek earlier. It had certainly been well worth the climb. However, we had to cut our picnic short as it began to rain, and the trails can become quite muddy and slippery when wet. It was much easier to descend then climb up and, I think, we were back in the rest house in under two hours.

On our way up we had stopped at the government-run Lalazar Wildlife Park which we were told housed a snow leopard. I was sceptical though because snow leopards are not found at these low altitudes. How then had they captured one? However, it transpired that there was indeed a snow leopard being held in a birdcage, which had been hurriedly readied for it when it first arrived in Lalazar in 2009.

How then did this rare and majestic animal end up in a birdcage? Politician Shahbaz Sharif's eldest son, Hamza Sharif, allegedly illegally acquired the snow leopard cub from a source in Central Asia. The Convention on International Trade in Endangered Species agreement bans the trade of endangered animals or their body parts. Hamza Sharif kept the cub in his house built atop a hill in Dunga Gali, and the baby snow leopard was reportedly well fed and cared for during the six months that it was there. When WWF-Pakistan learnt about this endangered animal being held in illegal captivity, they approached Hamza Sharif and convinced him to hand the animal over to a local zoo without any publicity. Once this was done, the male snow leopard was transferred to Lalazar Wildlife Park, which already housed a pair of male and female common leopards in a nearby purpose-built open enclosure ringed by a metal fence.

'Initially we wanted to put the male snow leopard in that enclosure, which is quite spacious, but he is capable of jumping to a height of almost 20 ft so he could have escaped. We had no other cage for him so we put him in here,' explained Mohd Riaz, the deputy ranger at Lalazar Wildlife Park. The cage which is enclosed completely with metal fencing was originally designed to house pheasants. It was sad to see this beautiful and rare animal being kept in such poor conditions. According to recent reports, the snow leopard is still there.

Monkeys continue to be found in abundance in the park and, due to their social nature, prefer hanging out by the road where tourists and passers-by feed them titbits. Sadly, they are sometimes killed by the local people or are trapped and sold to *bandar wallahs* (gypsies who earn money by teaching monkeys to dance and entertain). Illegal hunters also periodically sneak into the park to hunt and trap the pheasants; there are only twenty-four game wardens to safeguard 3,312 ha. The park also supports a 'healthy population' of the common leopard— perhaps its most notorious resident. In recent years, there have been many reports about leopards attacking the villagers and their livestock, even killing several people. In 2005, a common leopard became a man-eater, which led to 17 attacks and 7 deaths in the course of a year before it was shot dead.

The local community had begun retaliating by killing common leopards. According to Ashiq Ahmed Khan, then chief technical advisor to WWF-Pakistan, this was no solution as the leopard that knows their territory (usually marked by them over 50 sq. km) is a lesser threat than a new leopard moving in to claim the dead leopard's territory. 'Leopards are very territorial. The new leopard will not be familiar with the trails that the humans take, and will perceive them as a threat and invasion of his territory. Hence the new leopard will be even more dangerous,' he told me. Ashiq Ahmed's theory was confirmed when the local people later shot a leopard, only to have a new one move in and attack even more people. He said:

> We need to find a middle path. The people living here have to live in coexistence with the wildlife. Once the animals and the forests are gone, people will stop visiting the area. And more importantly, there will be no water as well. Due to all the cutting of trees by the locals outside the park area, 13 out of 26 streams have dried up. Without the forests there will be no water in the area.

Officials told me that there were around 20 to 30 leopards living in the Galiat region. According to Mohd Safdar, 'As the leopard population grows, they need more space. They are no corridors for them to leave the park so they end up in the village areas. We would like to extend the area of the park, and provide a safe corridor for the leopards.' The human-leopard conflict in Nathia Gali and adjacent areas was eventually addressed when WWF-Pakistan initiated the common leopard conservation project in 2010. They discovered that the unavailability of ungulates (wild goats) in the area led the leopards to rely on red fox, wild boar, birds, and small mammals for survival. Since even these were not sufficient in number, it resulted in increasing attacks of leopards on domestic livestock. According to WWF-Pakistan's wildlife expert Muhammad Waseem, 'We trained students about the importance of the common leopard for the ecosystem. For the first time in the history of Pakistan, a male common leopard was radio-collared by WWF-Pakistan on 1 September 2013 to monitor its behaviour in its natural habitat.'

Sungi had been working with the local communities in the area near the park at the time of my visit. Village organizations had been set up and a joint action committee appointed for forest protection. They had even begun disseminating environmental education, planting nurseries of alternative trees (for fuelwood), controlling land slippage, and opening nature clubs in the area. Indeed, the following day we had a delicious lunch up in the mountains near Tauheedabad with the local community. The children's nature clubs were there to welcome us with flowers and greeting cards. The project was also introducing non-timber forest products such as beekeeping for the production of honey. Their energy-efficient stoves, along with the introduction of the concept of kitchen gardens, proved to be quite popular as they enabled local women to plant vegetables for consumption. A schoolteacher from the area told me: 'This is a very positive approach. Self-reliance is the best—it is like a blank cheque that you can use anytime, anywhere. With self-reliance comes freedom.'

A meeting was set for the following day at the charming old hotel in which we were staying in Nathia Gali. Many of these issues were talked about at a round-table discussion with members of the local community, Sungi, and the staff of the national park. We listened to the debate about man versus leopard and local rights versus park rules, and the demands of the local people. Happily, the discussions were friendly and constructive. The recommendations that emerged in the end were quite sound: take only as much as you need from the forest, use the fuel-efficient stoves provided by Sungi or try to use LPG for cooking purposes as long as you can (there is no gas supply here), help protect and rejuvenate the forests, and please do not kill the leopards! Lastly, save your area for the sake of your water supply.

The Galiat area has seen both water shortages and flash floods in recent years. In November 2008, WWF-Pakistan began a project called Improving Sub-Watershed Management and Environmental Awareness in and around Ayubia National Park, which was funded by The Coca-Cola Foundation. The project worked with local communities to address the issues of clean water, sanitation, deforestation, overgrazing, pollution, and sustainable agriculture, and also established community-based organizations. Check dams have helped reduce sedimentation in the springs and the impacts of floods during heavy rainfall.[2] After the 2010 floods, WWF-Pakistan made gabion check dams that have also helped reduce flooding.

Afforestation was also introduced to improve forest cover and control soil erosion, and water storage tanks were provided to supply clean drinking water to the local community. The local women were sensitized about not cutting small trees and plants as this damages the ecosystem. To prevent the springs from being polluted, fences were erected around them and filtration plants were installed at local schools. Rainwater harvesting technology for many households was also provided.

According to Ibrahim Khan, a senior manager at WWF-Pakistan:

... the project has empowered communities to reduce their dependence on forests, protected natural springs from pollution, and raised awareness for a better future. A total of 118 ha have been planted with indigenous species of trees around Ayubia National Park.

In order to further reduce deforestation, more solar water heaters and fuel-efficient stoves were distributed among the local people. This has brought down the use of wood in traditional stoves by 30 per cent, which translates into reduced pressure on Ayubia's forests.

In view of these interventions, Ayubia National Park has managed to be successfully conserved over the years. Thousands of tourists continue to come every year to breathe in the fresh, pine-scented mountain air and enjoy the cool greenery of the forests. The local people are also happier as these projects have brought them many benefits. In the Galiat region where so much deforestation has taken place, everyone is grateful that Ayubia National Park, safeguarded by strict legislation, exists today.

Notes

1. *Ayubia National Park*. Website: <http://www.ayubianationalpark.com.pk>
2. 'Conserving the Environment & Watershed for Future Generations', *Coco-Cola Journey*. Website: <http://www.coca-colajourney.com.pk/stories/conserving-the-environment-watershed-for-future-generations>

6

Patriata: Last of Punjab's
Ravaged Forests

I REMEMBER PATRIATA CLEARLY FROM MY FREQUENT childhood trips to Murree when I would clamber on to the rickety chairlift to reach the top of the mountain, the highest point in Punjab. Due to mechanical failure, the chairlift has occasionally left people dangling helplessly; therefore, a ride on it entails an element of risk. Unlike Murree, the natural beauty of which has long vanished and destroyed through unbridled development, Patriata remained a haven: a dense forest which was free of any development and where it was possible to pleasurably saunter along gentle mountain trails. All this was about to change in 2003 as the Punjab government was planning an ambitious New Murree project, comprising a Rs 40 billion tourist resort and township project, spread over 4,000 acres of reserved forest land in Patriata. When the massive earthquake of 2005 struck the Kashmir region, the project was put into cold storage but only briefly.

'There's an active fault line that runs just below Patriata forest—here, you can see it on this map. This is restricted information, by the way. I managed to get a copy from the Turkish team that conducted a recent seismic survey in the region,' I was told by Arshad Abbasi while we sat in the IUCN (International Union for Conservation of Nature) office in Islamabad. Arshad is a writer, researcher, and a local resident of the area who was very concerned about the New Murree project. I had thought that the earthquake had effectively shelved the project; how could the construction of luxurious 'seven-star hotels' and an '18-hole golf course' be justified when just further north, nine districts of Pakistan had been devastated? Then there were the lessons learnt from the earthquake itself: most of the deaths that occurred in Pakistan's Azad Jammu and Kashmir region were due to landslides which swept away homes and schools located on heavily deforested land.

I was mistaken in imagining that we might have learnt something from the earthquake: the New Murree project may have come to a halt for a few months but the Punjab government was determined to continue with this project. It seemed that the prospect of 'developing' 4,111 acres of thick pine forest was just too alluring. Thankfully, the Supreme Court of Pakistan took *suo moto* action against the project in 2005 but the New Murree Development Authority (NMDA) gave the court a guarantee that the environment would not be damaged.[1] They were busy conducting a detailed environmental impact assessment (EIA) report that they planned to submit to the Supreme Court when I decided to visit the forest along with a fact-finding mission organized jointly by IUCN and WWF-Pakistan.

We set off from Islamabad on a bright April morning, packed into a small van full of environmentalists. We were headed to Patriata top, the site of the proposed New Murree Development project site, to see for ourselves what the NMDA was planning to do with the area. A public meeting had been arranged for us by the NMDA as we wished to hear the views of the local people. In the van were Arshad Abbasi, Ali Habib (then director general of WWF-Pakistan), Gul Najam

Jami (then with IUCN Pakistan), Helga Ahmed (a veteran environmentalist from Islamabad), Kamil Khan Mumtaz (the Lahore-based architect), his wife Khawar Mumtaz (who heads the women's NGO Shirkat Gah), Javed Jabbar (the former federal minister), and a few others. We decided to take the scenic route to Patriata from Barakho, off Murree Road, to enable us to see Simly Dam on the way.

The journey to Patriata took over two hours. Along the way, we were aghast to see large swathes of the countryside around Islamabad being 'developed' in the name of housing estates. There was construction everywhere, threatening all the small streams and nullahs that help raise the water table and eventually feed into reservoirs like Rawal Lake. On climbing up further, we were disturbed to see the unnecessary widening of the Koror-Patriata Road on both sides, bulldozers cutting into the cliffs, and trees being felled. I remember from my childhood, how heavily forested this area once had been.

Arshad Abbasi informed us that:

Massive deforestation began in the 1990s. Between 1990 and 2000, Pakistan lost an average of 41,100 ha of forest per year with an average annual deforestation rate of 1.63 per cent. Between 2000 and 2005, the rate increased to 2.02 per cent per annum. In all, between 1990 and 2005, Pakistan lost 24.7 per cent of its forest cover, or around 625,000 ha. The greatest victims were the coniferous forests of the lower Himalayan belt: Murree, Patriata, Galiyat, including the forests of AJK and the Kaghan-Naran Valley.

Today, according to WWF-Pakistan, Pakistan has a tree cover of only around 2.5 per cent which is considered to be amongst the lowest in the South Asian region.[2]

It would be natural to assume that the government would take steps to try to save what was left of these precious moist temperate forests. However, when we reached Patriata we realized that this logic somehow did not appeal to the NMDA. The British, on the other hand, had the sense to declare the forests of Murree and Patriata to be 'reserved forests' way back in 1887, to be protected not only for the region but also to harvest the monsoon rains and feed the river system below. The British actually called Patriata 'little England' after planting oak and chestnut trees on its mountainsides. Today, Patriata is awash with mature trees and its forested ridges offer magnificent views.

The NMDA was envisioning the construction of hotels, restaurants, golf courses, and shopping centres right in the midst of this healthy reserve forest, which is an important habitat for the common leopard as well as 14 other mammal species, 200 plant species, 146 bird species including rare pheasants, 22 reptiles, and 6 amphibians. This area is a key part of one of the finest remaining Himalayan temperate forest areas in the Punjab.[3] The forest guarantees high quality water with lower levels of sediments and pollutants for both the Simly and Mangla reservoirs.

Following our arrival in Patriata, we headed to the forest guest house where we were greeted by representatives of the local community (invited by the NMDA). They told us that they all welcomed the project because the government would build schools and hospitals, and encourage other development in this neglected *tehsil*. One by one they took turns to introduce themselves—schoolteachers, village elders, shopkeepers, landowners—and avowed that it was very important for the area's uplift for this project to go through.

'We need Sui gas, we need schools, we need roads, and

we need employment,' stated the local *nazim*. 'This project will be very beneficial for the area.' Everyone else seemed to concur: 'This project must not be stopped: our people are very poor; we need this development in this area. We want a town like New Murree.' It was pointed out that this project did not precisely envision a 'New Murree' but a high-end luxury resort for wealthy tourists. In response, they argued that they would still benefit from the rise in land prices and the development of schools and hospitals. We tried to explain to them that nowhere in the blueprint for the New Murree project was there any provision for schools and hospitals in the area. 'These are all basic facilities that you should be demanding from the government as your basic right anyhow,' Khawar Mumtaz pointed out.

A representative from the NMDA stated that that once they had set up their office in Patriata, the theft of forest trees had ceased. 'If we don't start this project, I can guarantee that in twenty years all the trees will be gone—all stolen by the local community.' He admitted, however, that no large timber mafia operated in the area. Azmat Ali Ranjha, then the director general of the NMDA, added that they planned to plant more trees in the project area on around 467 acres of land. 'We plan to have hiking trails, a wildlife sanctuary, a bird sanctuary; and will only use the existing tracks so that all these environmentally friendly activities can be enjoyed by entire families who can come here to relax.' He pointed out there were hardly any recreational activities for people and the New Murree project would provide local tourists with all the necessary facilities in an environmentally friendly manner. The question remained: how were golf courses, chalets, hotels, shops, and a community centre going to be constructed without ravaging the fragile forest environment?

Furthermore, who was going to guarantee that thousands of tourists each year would not further degrade the forest? 'The NMDA will have full authority—we shall have our own laws, our own government,' said Azmat Ranjha. On hearing this, Ali Habib jocularly inquired whether they would also be applying for their own seat at the United Nations.

Ali Habib then proceeded to ask members of the local community who had been assembled whether there was anyone who opposed the project. Thereupon Hafiz Saeed Ahmad, a member of a local NGO called the Rural Areas Development Organization, raised his hand. He said:

> The local people are not involved in the NMDA's decisions. This is not a participatory process. The rights of the people are not being respected, let alone the rights of the forest. We were also badly affected by the recent earthquake but none of us have received any compensation from the government. So what hopes can we have that they will bring us development through this project? Even the political representatives of this area are opposing this project. The political parties have not been involved—only the bureaucrats. They bulldozed this bill through the Punjab Assembly to create the NMDA.

The other local people who had gathered tried to prevent Hafiz Saeed from continuing further but, to the credit of the NMDA officials, they allowed him to complete his intervention. Later, Naveed Nazir, a burka-clad member of the District Council and a fiery opponent of the project, told us: 'Our identity will be destroyed. We shall become displaced persons like those affected by the Tarbela Dam. We our selling ourselves and the forests planted by our forefathers so cheaply.'

We decided to go for a walkabout and climbed up to Patriata

top. To our delight we saw wild horses galloping through the forest between thick, hundreds of years old deodar trees. It was an incredible sight to behold and we stayed at the top for a while, breathing in the clean mountain air and marvelling at the towering trees that formed a green canopy above us. The late afternoon sun filtered in through the treetops forming filigree patterns along the pine-covered forest floor.

On our way down, Azmat Ranjha pointed to a nearby ridge and informed us that a tunnel would be built through it to facilitate the traffic moving to the surrounding villages. This would prevent the local people from even having access to the road that currently cuts across the project site to the villages beyond. This new addition to the plan confirmed our fears that the project managers were not sharing the details with the local people who were completely in the dark about what was in store for them.

'Why are you opposing this project?' some people asked me as we walked back to our van. 'Because you will be the first people to be displaced by this project; it is being built for rich people from elsewhere in the country, not for your benefit.' I tried to explain the situation to them. 'I wish you would look at the blueprint for the project and see for yourself that this is not another Murree but a luxury resort: there will be no schools or hospitals built for any of you.' One of them scrambled off, only to return with a piece of paper listing in bullet points all the salient features of the project: boutique hotels, five-star hotels, Mall walk, 18-hole golf course, etc. They pointed to a section that read: 'Spa and health facilities', and said, 'See, *baji*, here is our hospital right here.' I didn't know whether to laugh or cry.

Finally in 2008, after an extended media and legal campaign led by a coalition of environmental organizations

and with the coming of the new government of Shahbaz
Sharif in Punjab, the proposed tourism development project
in Patriata was put to a stop. Pakistan's Supreme Court not
only formalized the new government's dissolution of the New
Murree Development Project but also ruled that there should
be no future projects of that kind in the area. WWF-Pakistan,
which had become a party to the case, was delighted that an
important watershed and habitat had been saved.

Iftikhar Muhammad Chaudhry, chief justice of the Supreme
Court at the time, must be credited for using his authority (in
September 2005) to halt the project pending a judicial review
of the proposal. In 2009, the bench observed:

> It is noteworthy that all over the world national parks are
> developed to preserve flora and fauna facing [the] threat
> of extinction in the wake of modern-day life development
> projects including [the] mushroom growth of housing projects,
> recreational facilities, etc. The need is to sensitize the general
> public to the fundamentals of sustainable development so as
> to achieve the goal of a healthy environment, not only for
> the present population but also for the future generations.
> The concerned agencies of the Government, including
> Environmental Protection Agencies at different levels, have a
> heavy onus to discharge in this regard. The Government of the
> Punjab, considering the environmental hazard posed by the
> New Murree Development Project, has taken a right decision
> in disbanding the same.[4]

The Punjab government shelved the controversial New
Murree Development Project and returned the 4,111 acres
of land acquired for the purpose to the Forest Department.
The Supreme Court's decision was hailed as a victory for all

the environmentalists in Pakistan who had banded together to ensure the project was stopped. Pakistan was saved from the nightmare scenario of thousands of trees being felled, the fragile ecosystem of Patriata being damaged beyond repair, and wildlife habitats destroyed overnight merely to develop a playground for the elite.

Notes

1. 'Environmental woes made project unfeasible: report: New Murree Development Authority', *Dawn.com* (June 2008).
2. Syed Muhammad Abubaker, 'Save Falling Trees', *WWF-Pakistan's Blog*. Website: <http://wwf.org.pk/blog/2015/02/03/save-falling-trees/>
3. 'WWF celebrates saving of Himalayan forest and not so Common Leopard, *Panda News* (Aug. 2009). Website: <http://wwf.panda.org/wwf_news/?uNewsID=172481>
4. 'Excerpt: Environmental Activism and the Law', *Dawn.com* (Apr. 2015).

7

Chilgozas: Black Gold of Suleiman Range

NEVER HAVING SEEN A CHILGOZA TREE BEFORE, I WAS QUITE excited to find myself on my way to Zhob district from Dera Ismail Khan in 2006. Our jeep was headed to the Suleiman Range on the border between Balochistan and Khyber Pakhtunkhwa (this was in the days prior to the spread of the Baloch insurgency when the journey to Balochistan was considered quite safe for outsiders like me). The Suleiman Range is an extension of the Hindu Kush Mountains into Balochistan. The steep mountainous terrain, the hot and dry summers, and the cold winters all combine to provide for the world's largest pure stands of chilgoza.[1] The Chilgoza forests had come under threat because they were being cut by local tribesmen until WWF-Pakistan intervened. Today the chilgoza nuts produced by these forests are called 'black gold' because of the high prices they fetch in international markets. The local people too have become aware of the value of these nuts.

I was travelling with a driver and a local Shirani tribesman (the Pakhtun Shirani tribe lives in this area on both sides of the provincial border). Ahmed was to be my guide because we were crossing into tribal areas where the government's writ does not hold, although I do think that the tribal areas needlessly get a bad reputation in Pakistan. Personally, I have never met more hospitable people than those who inhabit these supposedly wild regions. However, it is true that feuds are commonplace and almost every man carries a gun.

The drive through the Suleiman Mountains kept me riveted to the car window with my camera. Unlike the tempestuous Himalayas with their rolling boulders, seismic activity, and perennial landslides; the Suleiman Mountains are older, settled, and haunting in their bare ruggedness. Halfway through the four-hour drive, we came across the Kuchis (Afghan nomads who seasonally migrate to Pakistan); hundreds of them were camped near the dry riverbeds. They were wintering in Pakistan having migrated down south, as they have for centuries, with their homes laden on camels and trucks. Their women were colourfully attired with their silver jewellery and long, Pathan dresses, but kept away from the road. The men were striking with their tall and muscular physiques, wearing enormous black turbans, their weather-beaten faces expressing character. 'Will you send us the pictures?' they quipped as I took photo after photo. They were shepherding their flocks of sheep on the road, walking with such pride and dignity that I wanted to stop and visit their colourful camps. Unfortunately, it was getting late and we had to reach Zhob before dark.

'They are being friendly with you, but otherwise they tend to keep to themselves and don't really like outsiders,' Ahmed informed me, a little alarmed at the prospect of having to take me into one of their tents. He explained that the Kuchis ran a lucrative smuggling operation trading in guns and armaments on both sides of the border. Nevertheless, I thought it was wonderful that these people were able to enjoy the freedom of a nomadic lifestyle in our corner of the world even in the twenty-first century. For centuries, this tribe has wandered across the deserts and mountain passes of Afghanistan and Pakistan with their herds of sheep and camels. A change has now been occurring however. I recently read that the Kuchis are putting down roots and building permanent homes and villages in Afghanistan. Decades of war and turmoil in Afghanistan and a long period of drought have finally taken a toll on them.[2]

We arrived in Zhob just as the sun was setting. Regretfully, I was unable to see much of this dusty little town, named Fort Sandeman by the British, because early the following morning we penetrated deep into the Suleiman Range where WWF-Pakistan had set up a small field camp. Happily, I did get a chance to see the Milky Way at night. As we sat in the garden of the WWF office where I was staying, the electricity went off and I looked above only to be left stunned. I had forgotten what stars look like in smog-beset Lahore. Here in Zhob, they appeared to be hovering just above you. My hosts, Bazmir Khan and Imran Khan who then both worked for WWF, left me to my quiet reverie gazing up at the stars and recollecting my favourite saying: 'Ideals are like stars—we may never reach them but like the ancient mariners, we can chart our course by them.'

The next day was uneventful with Bazmir Khan making plans for an expedition up the mountains to see the chilgoza trees. The growing demand for chilgoza nuts in places like Dubai, Muscat, Jeddah, London, and Jerusalem makes for high prices. I even met some chilgoza merchants visiting from Dera Ismail Khan who told me that the nuts from the Suleiman Range are probably the tastiest in the world. Haji Qasim Khan, a merchant from Dera Ismail Khan, said: 'The big market is in Lahore. We send 80 per cent of the chilgoza nuts to Lahore where many are peeled and then exported abroad.' WWF-Pakistan wanted to eliminate the presence of middlemen so that the merchants could buy directly from the local tribesmen and thus get better prices for their chilgozas.

We spent the night at the field office, which was merely a cluster of mud and thatch rooms with a protective stonewall, perched atop a ridge; below us were agricultural fields owned by the local villagers. The project has also helped reduce the villagers' dependence on the Chilgoza forests by improving agricultural productivity (as apparent by an increase in the variety of crops grown). Earlier, the water channels had almost stopped functioning because of sundry tribal feuds that are common in this region. The project helped settle disputes, and now the villagers can grow onions and other vegetables which fetch good prices in the markets. The local people even grow fruits such as apples, grapes, and pomegranates. The flat valleys had good grazing land. At higher elevations, there were olive and mixed scrub forests (including pistachio and acacia). This area is evidently quite rich in wild fruit, nuts, medicinal plants, and mushrooms.

Nearby, a stream flowed down from the mountains encircling the valley. The green fields and the austere, blue-grey mountains provided for a dramatic contrast as the sun set. The stone mountains looked forbidding and, in retrospect, I should have paid more attention to Bazmir's enquiries about my capacity to climb up the mountains in my jogging shoes. I was told later that these are cruel mountains and I certainly discovered this the hard way.

I had wanted to see the Chilgoza forests up close, and in order to do that I had no choice but to climb up the mountains. The chilgoza trees are only found at higher elevations between 2,133–3,353 m. In total, approximately 26,000 ha of forest remain in the Suleiman Range. Accompanying us was a local tribesman, Yar Mohammad, a member of the Shirani tribe who was a striking presence in his grey turban, a Kalashnikov slung across his shoulder. Yar Mohammad lived in the nearby village of Ahmadi Daragh. Luckily, we had him as our guide on our way up and could turn to him for directions when the terrain became extremely hostile.

We set off at 5 a.m. with donkeys laden with food and knapsacks leading the way. After what seemed like hours trekking through a stony riverbed (dangerous during the rainy season when hill torrents can come cascading down the mountain, sweeping away rocks and anything else in their way with their sheer force), we finally started climbing. By now, the sun was blazing away although it was mid-September. Fortunately, the higher we climbed, the cooler it became. Since there is no water available at the top of the mountain, we carried canisters which we filled from a freshwater spring on our way up.

Imagine our surprise then to discover that we had to stand aside to give way to camels descending the narrow mountain track almost half way up. 'Yes, the camels go all the way up—they are less in number now. Earlier the local people would use them extensively to bring down timber from the mountains,' explained Bazmir who was also accompanying us. 'We have constituted a community-based Chilgoza Forest Conservation Committee and they are enforcing a fine of Rs 10,000 on any villager cutting a green chilgoza tree in their area. People realize now that these trees are precious. We want them to get the proper price for the chilgoza nuts,' elaborated Bazmir. Consequently, the local people had begun cutting fewer trees in respect of the ban imposed. They would only bring down the older trees which had dried up to either sell or use as fuelwood.

The local communities had become wise to the fact that in a good year, the value of nuts produced by a single chilgoza tree is higher than the timber value of an entire tree. 'Earlier, they would extract the chilgozas too early and damage the trees as well. Now we have taught them the best time for extraction, which is early September, and have given them special equipment to cut the cones,' said Bazmir.

Although this area is a part of Balochistan, there is no Sui gas in Zhob district. The people of this tribal area are very peaceful—there is little crime, and most are members of a single tribe, the Shirani. They own almost 1,200 sq. km of pure Chilgoza forest.

As the track began getting narrower and steeper, we came across our first chilgoza trees, so delicate and spindly. They were like mini-pine trees but this of course was new growth and we had yet to reach the dense forests above. We took a break for tea, climbing up to a summer camp nearby. By now we were exhausted but had yet to reach the top. Near the camp, hundreds of pine cones were littered by the makeshift shelter. It seemed like we had just missed the harvesting because the local villagers had already extracted all the chilgoza nuts and then buried them in safe hiding places. 'They will come for them in a week or so—they are letting them dry out a little,' explained Yar Mohammed. He was also able to locate one or two hiding places and showed us how the nuts are extracted from the cones. It appeared to be back-breaking work, not to speak of the climb up and back down the mountain.

Our final push to the top of the mountain, called Pazi, was hampered by a landslide which had swept away the track. We had to now use both hands and feet to scramble up, and Yar Mohammed lent me his unravelled turban to use as a makeshift rope. After around half an hour of hard climbing, we finally reached the top. There was plenty of new growth of chilgoza trees which was an encouraging sight. 'In another twenty years, this will be a thick forest,' said Bazmir. 'The average chilgoza tree lives up to around a hundred years and gives fruit every year.'

Just across us, on the other side of the valley, was the famous Takht-e-Suleiman (probably around 1,000 ft higher). There is a shrine at the top which is said to be the tomb of the founder of the Pukhtun race. On being informed that camels can take people right to the top, we wondered why we hadn't brought them along on our trip. Then we were told that the nasty smell along one part of the trek was that of a dead camel; so they do after all occasionally slip and fall when negotiating these dangerous mountain tracks.

The sight of the dense Chilgoza forest on one side of the mountaintop, which was virtually shaped like a plateau rather than a pointed top, caused me to forget the arduous trek up the mountain. It was a truly splendid sight: a forest so dense that one could not even see the mountain's slopes. In between the chilgoza trees were blue pine trees. The blue pine is a beautiful tree, almost like a Christmas tree, but even prettier with its blue-green colouring and dense pines needles. We took lots of photographs and reluctantly turned away as it was time to head back down before dark. There are no villages or settlements on the mountain. Once the sun sets, it gets very cold up there and, besides, there is no water at this height. The local villagers who are tough mountain people can easily trek up and down in a few hours. If need be, they spend a few nights there during harvesting season, lighting a huge bonfire to keep themselves warm at night.

We had the option of spending the night at the campsite but by now the mountain was becoming inhospitable. It was very painful coming down the mountainside in joggers (after all, one needs proper hiking boots for this kind of trek) and the stones were causing me to stumble. Regardless, the pain made me forget my exhaustion and I was determined not to be defeated by this mountain so I hobbled down.

By the time we reached the riverbed, it was already dark but fortunately we had torches to light our way. From the distance we spotted the jeep that was waiting to take us back to Zhob. 'This is a hard mountain,' Yar Mohammed told us. 'Not many visitors can go up and down in one day!' Indeed, the driver told me that he had said many prayers for me while waiting with the jeep. When I got back to the rest house and took off my shoes, I realized to my horror that my toenails had been bleeding. Every muscle in my body was aching, so soon after dinner I went off to sleep.

Yet for the local people this is an everyday reality. They go up and down these mountains at least once a week to fetch fuelwood and timber. During harvesting season in early September, they go up almost every day. I came away from the mountain with renewed respect for these tough, dignified tribesmen. With the right guidance and encouragement, they have a prosperous future ahead of them. The market for chilgozas is not only going to get bigger but as demand grows, so will the prices soar. Already, each family in the area makes over Rs 50,000 a year from selling the chilgoza nuts grown in these community-owned forests. Today the market price of peeled chilgoza nuts in Lahore is around Rs 1,000 per kg.

To improve the livelihoods of the communities living around the Chilgoza forests, WWF-Pakistan, in collaboration with Agribusiness Support Fund (ASF), took the initiative to assist the local people in processing and marketing chilgoza nuts. People were organized under the Chilghoza Forest Association (CFA) of the Suleiman Range. From 2012 to 2014, WWF-Pakistan provided support in the form of training. Pre- and post-harvest training to chilgoza nuts collectors was conducted as well as the sowing of chilgoza seeds in empty areas. Nurseries of forest and fruit plants were also established, and to reduce pressure on forests for fuelwood, fuel-efficient stoves and solar water heaters were distributed to the communities. The project also set up chilgoza, olive, and wild pistachio processing units. By organizing the villagers and teaching them how to manage their precious forests and market their produce, the project not only helped to save the Chilgoza forests but also brought the people closer together as a harmonious and peaceful community.

Notes

1. 'Chilghoza forest', *WWF-Pakistan*. Website: <http://www.wwfpak.org/ecoregions/ChilgozaForest.php>
2. Robert J. Galbraith, *The Last of the Afghan Nomads Enter the Modern World* (Apr. 2013). Website: <http://www.robertgalbraith.com/the-last-of-the-afghan-nomads-enter-the-modern-world-4/>

8

Living Fossils: Juniper Forests of Ziarat

I LOVE FORESTS, AND THIS IS PARTICULARLY SO BECAUSE WE have so few left in Pakistan. Most of the country is arid and treeless so visiting a dense forest is like coming across a waterbody at an oasis. I had wanted to go to Ziarat to see its famous Juniper forests for a number of years, and eventually got an opportunity to do so in 2004. Ziarat is the largest contiguous natural Juniper forest in Pakistan spanning 100,000 ha. It is also the second oldest, trailing behind the one in California. A three-hour drive from Quetta, Ziarat is named after the famous shrine which is located there of a local saint called Kharwari Baba. He is buried around 10 km from Ziarat and is said to have blessed the valley. The town is also famous for the elegant Residency where the founder of Pakistan, Quaid-i-Azam Muhammad Ali Jinnah, spent his last days.

The British developed Ziarat as a hill station and adopted it as their summer headquarters in 1882. Indeed, the Quaid-i-Azam Residency was the old residence of the agent to the British governor general and was built in 1892. It was the first place I went to visit when I reached Ziarat. We had to climb uphill to reach the metal gates that led to the lush green lawns with graceful chinar trees and flower gardens. The double-storey house with its wide verandahs and wooden flooring was even more picturesque than it appeared in photographs. The Ziarat Residency was indeed a magical building soaked in history: the estate had long been declared a national monument and heritage site, and it was well preserved. Inside, I climbed up a creaking, wooden staircase and peered into Jinnah's austere bedroom with its graceful wooden desk and tidy bed. On the walls were black and white images of the Quaid with his colleagues from Balochistan.

In his book, *With the Quaid-i-Azam during his Last Days* (2011), Dr Ilahi Bakhsh, Jinnah's personal physician, has provided a detailed account of the Quaid's treatment at Ziarat and Quetta, and his final journey to Karachi.[1] According to Dr Ilahi Bakhsh, the Quaid's condition actually began improving in Ziarat but when he was brought down to Quetta for some tests, he developed an infection. Dr Ilahi Bakhsh and other doctors informed Fatima Jinnah that unless there was a miracle, there was no chance of the Quaid-i-Azam surviving for more than a day or two. Arrangements were made to fly him to Karachi. When Jinnah finally arrived back in Karachi on the day which would prove to be his last—11 September 1948—a rickety ambulance was sent to receive him at the Mauripur Aerodrome and that too without a nurse. The ambulance broke down on the way to the city which must have hastened his end in the oppressive heat of September. His devoted sister, Fatima Jinnah, was with him. They had to wait for over an hour for another ambulance to arrive and Jinnah passed away later that evening.

I was happy to see that the Quaid's Residency had remained so well maintained; some ascribing this to the juniper wood that was used in its construction. The dry, cool air on the juniper-clad hillside was also good for the Quaid when he spent his last days battling lung cancer and trying to cling on to life. The site was especially chosen for the ailing leader because Ziarat has a very salubrious climate, especially in summer. Besides, Jinnah had a great personal fondness for the place. I could just picture him sitting on the verandah, staring into the lush forest that surrounds the Residency and the valley below, contemplating the future. He probably was aware that he was going well before his time.

Unfortunately, several years after my visit this treasured national monument was burnt to ashes when militants from the outlawed Baloch Liberation Army (BLA) carried out bomb attacks on the Ziarat Residency on 15 June 2013. So intense was the terrorist attack that the flames took days to extinguish. Most of the original wood had been burnt leaving only the metal and stone foundation standing. The Government of Balochistan, however, had it restored to its original form after months of intense restoration efforts by a team of well-known experts at a cost of Rs 140,000,000. The renovated building was inaugurated by the prime minister of Pakistan on 14 August 2014.[2] I have not been back since to see it but in photographs it seems to very closely resemble the original.

Ziarat's juniper species are of global significance because they survive for long and grow quite slowly. The species are ranked as being among the oldest living trees on earth. The age of a mature tree exceeds 4,000 to 5,000 years, and because of this they are locally termed 'living fossils'. The gnarled old branches of these ancient trees is a remarkable sight and a drive through a Juniper forest an unforgettable experience. On our trip to Kharwari Baba's shrine, we drove through the thick Juniper forest. Over there the rocky cliffs were a wondrous sight to behold with these amazing trees somehow emerging from the crags.

Ziarat literally means pilgrimage. As noted earlier, the town derived its name from the shrine located nearby of the famous Muslim saint Mian Abdul Hakim, popularly known as Kharwari Baba. The shrine is situated in the valley south of Ziarat town. According to legend, the Sufi saint came here from Kandahar in Afghanistan. He opposed the high-handedness of the local Afghan king and was obliged to leave his native town and migrate to Ziarat. When he reached this valley, he took up his abode on a hilltop and prayed for the land saying: 'This place shall flourish.' Thereafter water began oozing out of the spot; it continues to flow and is regarded as being holy.

I remember spending the rest of that happy afternoon in the Juniper forest, looking at all the different species growing on the forest floor. The floor was in fact covered in emerald green foliage that contained many hidden treasures in the form of plant species of medicinal significance. I was told that of the 54 available species, over 50 per cent are of medicinal/ethno-botanic value and are traditionally used by the local communities to treat a variety of medical conditions. They could potentially be of use to pharmaceutical companies. Already there is a herb called *Ephedra sinica* that is found in abundance here, from which a chemical called ephedrine is extracted. This is an important constituent of a number of medicines especially cough syrups.

Unfortunately, this fragile ecosystem is under tremendous pressure. The local population of Ziarat is growing and when we visited the area, people were cutting the juniper trees to use as fuelwood in the chilling winters. The town itself is unimpressive: a nondescript cluster of rundown hotels, houses, and government buildings. In 2004, Sui gas was finally provided to Ziarat by the government although not everyone has a gas connection. The gas pipeline only provides gas to the main town of Ziarat and most villages in the valley do not have access to it. During the winter, the pressure in the pipes drops drastically, leaving the local people without gas for weeks at a time. In that season, temperatures in Ziarat plunge to 18° below 0 so, in the absence of gas, people have no alternative other than to chop trees for heating and cooking. The local forest department complains that it lacks the resources in terms of manpower and vehicles to police the forests.

There has also been less snowfall in the area so many people do not migrate in winters as they once used to. They now stay on in Ziarat; their numbers are increasing as the hill station has been declared a district headquarters.

The local people living in the forest also have other uses for the juniper wood. They employ branches from the trees to fence their domestic animals and thatch their roofs (as the wood is impermeable to water). Their livestock devours the forest's ground cover as well which includes the invaluable medicinal herbs. Although they don't consume juniper trees, they trample over young seedlings. As it is, juniper is an extremely slow growing species gaining only around an inch in height each year. A hundred years old tree is considered to be very young. Ziarat's Juniper forest lacks generation and it was extremely sad to see its decline.

During our visit, we witnessed many of the older trees falling victim to diseases for which there are no cures. The trees simply had to be cut down to prevent diseases from spreading. The government did not have the resources to protect this large forest; therefore, an agreement was signed at that time with the UNDP which sought to involve communities and the government in conserving and managing the Juniper forests in Balochistan. The IUCN (International Union for Conservation of Nature) was also involved as the implementing agency. The project entailed educating the local communities on the value of the juniper tree to incentivize them to protect the forest.

These ancient forests were added to the World Network of Biosphere Reserves in 2013. Biosphere reserves are ecosystems which are recognized internationally under UNESCO's Man and the Biosphere Programme (MAB). They are established to promote a balanced relationship between humans and the biosphere. Biosphere reserves—of which there are currently 610 around the world—are designated by the International Coordinating Council of the MAB Programme at the request of the state concerned. The Juniper forests of Ziarat is the second biosphere reserve in Pakistan to be added to the list. The first was Lal Suhanra National Park situated in Bahawalpur district.[3]

'The inclusion of the Juniper forest of Ziarat as one among the World Network of Biosphere Reserves is yet another step towards recognition of Pakistan's natural sites of international significance at a global level, which is a matter of great honour and pride,' says Mahmood Akhtar Cheema, country representative of IUCN Pakistan. The process of getting the Ziarat Juniper forests included as a biosphere reserve was initiated by IUCN Pakistan in 2010 under its UNDP funded project in collaboration with the Balochistan Forest and Wildlife Department. To meet the criteria, a management plan was formulated and approved by the Government of Balochistan. Consultations were also held with local communities and other relevant government departments. According to UNESCO, a biosphere reserve is not only a habitat for vegetation but also for animals, insects, and man as a single holistic nature reserve.

In early 2017, UNESCO also accepted Pakistan's recommendation to include eight Pakistani sites, including the Ziarat Juniper forests, in its list of World Heritage Sites. Pakistan's archaeology and museums department will soon start the documentation process which will be completed within two to three years.

Recently the Balochistan Provincial Assembly passed a resolution declaring Ziarat to be a tourist resort and vowed to refurbish the hill station where Quaid-i-Azam had spent his last days. A plan is now underway to develop Ziarat and investors have been urged to set up good rest houses and hotels so that more tourists can visit the site. We can only hope that this development takes a sustainable form and that Ziarat does not become another Murree which has been utterly ruined by an overgrowth of concrete buildings.

The Juniper forests of Ziarat also provide a habitat for several endangered wildlife species like the Suleiman markhor, urial, black bear, and wolf. The forests are also home to several species of migratory birds. In addition, the forests help recharge the aquifers which support the livelihoods of the local communities.[4]

Above all, the Juniper forests of Balochistan are an ecological and cultural treasure of Pakistan and the world, and therefore their preservation and well-being is imperative.

Notes

1. Ilahi Bakhsh, *With the Quaid-i-Azam during his last days* (Karachi: Oxford University Press, 2011).

2. Syed Ali Shah, 'Rehabilitated Ziarat Residency to be Inaugurated on August 14th', *Dawn.com* (May 2014).

3. 'Juniper Forests, Ziarat', *WWF-Pakistan*. Website: <http://www.wwfpak. org/ecoregions/JuniperForests.php>

4. 'Ziarat Juniper Forest', *Unesco.org*. Website: <http://www.unesco.org/new/ en/natural-sciences/environment/ecological-sciences/biosphere-reserves/ asia-and-the-pacific/pakistan/ziarat-juniper-forest/>

9

Torghar: Markhor Heroes
of Balochistan

TO REACH THE TOBA KAKAR RANGE ON THE BORDER OF Afghanistan, I had to undertake a gruelling nine-hour jeep drive from Quetta, the capital of Balochistan. In these mountains, the local people still observe centuries old tribal customs and style of life: they live in tents made of black goat hair, and move around the region in a seasonal cycle to feed their sheep and goats. I was travelling there as a guest of Sardar Naseer Tareen who jokingly told me that we would be staying at the local 'Holiday Inn'. When we arrived there was only a cluster of mud huts (with no flushing toilets) which he had built for guests on a small ridge. This was to be our 'hotel' for the next couple of days. There are only four settlements in Torghar, the largest being Tanishpa, where some cultivable land and water from springs is available for the establishment of orchards and growing crops. The tribesmen are renowned as much for their hospitality as they are for their fierce pride, and I observed that many were carrying guns (I don't think I even saw another woman for the entire duration of my stay).

The Torghar landscape is rocky and barren, and plants only come to life when there is rainfall. The vegetation consists largely of scrub grass and wild pistachio and juniper trees. Near the streams, however, there is an abundance of plant life and small animals living among the large boulders. Wild tulips can often be spotted and over 80 species of birds have been recorded.[1] Sardar Naseer has also ascertained that the hawfinch is not a seasonal visitor to the region but that it actually breeds in Torghar. There has also been the rediscovery of the Afghan mole vole, a small rodent that was thought to have vanished. In addition, scientists have discovered two new species of rock lizards, one of which has been named after Sardar Naseer!

Sardar Naseer calls himself an 'accidental environmentalist' because by training he is a film-maker. He became involved with this region during the making of a film on the wildlife of Balochistan back in 1983. 'I was not a conservationist ... my interest lay in film-making. I was pushed into the pool, in a manner of speaking,' he recalled in slightly American accented English as we relaxed outside his 'Holiday Inn' sipping hot tea after the arduous jeep ride from Kila Saifullah, the nearest town. Sardar Naseer comes from a town called Pishin near Quetta. At the time I met him, he was already in his late sixties. He had and still has all the sophistication of a globetrotter who is equally at home in both the East and the West. To this day he remains a bachelor, saying 'that is why I have all the time in the world for my work!'

He attended Lahore's prestigious Government College from 1953–58 before leaving for the US. 'I was going to study International Relations ... but instead I switched to Communication and got admitted to the California Institute of Arts,' he said with a broad smile. In the 1960s, CalArts was a highly selective institution and only a few who showed promise of outstanding talent were granted admission. While Sardar Naseer thoroughly enjoyed his studies in film-making, back in Balochistan his conservative family was upset by their son's unconventional decision to pursue the arts. He nevertheless completed his training and decided to stay on in the US. At the time his father was the Sardar (or chieftain) of the Tareen tribe, as is his nephew today (all the sons can use the title of 'Sardar'). Sardar Naseer's ancestral home is an old fort located about 30 km north of Pishin, and is still inhabited by members of his extended family.

'I never really had the intention of settling in the US, and was forever running back and forth, and one day I said "let's go home",' he recalled. After living in the USA for twenty-four years, Sardar Naseer determined to reconnect with his tribal roots. In 1983, when he returned to Pakistan, he thought of making a feature-length film on Balochistan's cultural heritage for the overseas market. The Government of Balochistan asked him to first make a short film on the wildlife of Balochistan. He soon discovered that the animals they listed only existed in the files of the provincial government's wildlife department.

Traditionally, hunting had been practised in Balochistan's tribal areas but it was only during the Afghan war in the 1970s that the wildlife of the province began being exterminated. Sardar Naseer concentrated his filming efforts around the area known as Torghar in the Provincially Administered Tribal Areas of north-eastern Balochistan. Called the 'Black Mountain', Torghar consists of a series of dark-coloured upturned ridges which are approximately 90 km long and the altitude varies between 2,500–3,300 m. These ridges are home to the last pockets of Afghan urial and Suleiman markhor in the world; very few of the latter being found outside Pakistan. The Suleiman markhor is the straight-horned species that exists only in Balochistan and Khyber Pakhtunkhwa. As the total world population of this unique animal is found largely within Pakistan, the country has a special responsibility to ensure its future survival.

Sardar Naseer realized that the wildlife of Torghar would soon disappear if no protective measures were immediately put in place. Aside from indiscriminate hunting, massive deforestation and land erosion were making survival difficult for the wildlife of the area. In the town of Qila Saifullah, Sardar Naseer discussed the issue of the dwindling wildlife in the Toba Jogezai Range with the late Nawab Taimur Shah Jogezai who was then chief of the Kakar tribe. Sardar Naseer Tareen is related to the Jogezais through several marriages between the two families, including his sister's marriage to the late Nawab Taimur Shah's brother. Nawab Taimur Shah told Sardar Naseer about his fruitless efforts to petition the provincial government to provide gamekeepers who would protect the wildlife in his area. They both agreed that some action must be taken before it was too late. In 1984, Sardar Naseer initiated a conservation programme in Torghar with the support of Nawab Taimur's son, Nawabzada Mahboob Jogezai, who he says became a 'fanatic conservationist'.

While we dined later inside the rest house, a cold wind began blowing outside. A full moon rose up from behind the mountains and lit up the lunar-like landscape. Somewhere along these ridges were the markhor. I was all set to go for a moonlit climb to look for a herd frolicking in the cool night breeze but was politely dissuaded. 'You'll find out what it is like tomorrow morning … it's not that easy to spot a markhor,' I was told. I remembered these words later when I found myself struggling up a precipitous mountain ridge, with the loose gravel and rocks rolling beneath my feet and my hands grasping for the shrubs that grow high up on these cliffs.

I was awoken at 6 a.m. to look for markhor and was told

2.1 Shandur polo final—half-time break (*Photo by Author*)

3.1 The bridges in Chitral washed away by the floods (*Photo by Syed Harir Shah*)

3.2 The damaged PTDC Motel in Kalash Valley (*Photo by Author*)

3.3 Kalash girls in front of their intact homes built on higher ground (*Photo by Author*)

3.4 The children of Rumbur Valley traumatized by the floods (*Photo by Author*)

3.5 Kalash children running in front of Bumburet nullah that flooded causing much damage (*Photo by Author*)

4.1 Freshly cut tree in Kohistan (*Photo by Author*)

4.2 Picturesque Palas Valley in Kohistan (*Photo by Author*)

4.3 Piles of timber lying on the Karakoram Highway near Pattan (*Photo by Author*)

4.4 The pristine forests of Palas Valley (*Photo by Saleemullah*)

4.5 A young Kohistani boy (*Photo by Sohail Siddiqui*)

5.1 Pipeline walk at Ayubia National Park (I) (*Photo by Author*)

5.2 Pipeline walk at Ayubia National Park (II) (*Photo by Author*)

7.1 Checking for chilgozas in the pine cone (*Photo by Author*)

7.2 Taking out fresh chilgoza (*Photo by Author*)

8.1 Ancient trees of the Juniper forest, Ziarat (*Photo by Author*)

8.2 Ziarat Residency before it was burnt down and rebuilt (*Photo by Author*)

9.1 The author in Torghar Hills (*Photo taken with author's camera*)

9.2 Torghar Hills known as the 'Black Mountain' (*Photo by Author*)

9.3 Straight-horned Suleiman markhor in Torghar (*Photo by WWF-Pakistan*)

10.1 Biogas digester in Soan Valley (*Photo by Author*)

10.2 Check dam in Soan Valley built to recharge groundwater (*Photo by Author*)

11.1 The Lansdowne Bridge in Indus Dolphin Reserve, Sukkur (*Photo by Author*)

11.2 Tombs of seven virgins in Indus Dolphin Reserve, Sukkur (I) (*Photo by Author*)

11.3 Tombs of seven virgins in Indus Dolphin Reserve, Sukkur (II) (*Photo by Author*)

11.4 Indus dolphin rescue (*Photo by Jameel Ahmed*)

11.5 A fisherman holds a dolphin during a rescue near the Indus Dolphin Reserve (*Photo by Hussain Bux Bhagat*)

12.1 A Chotiari fisherman (*Photo by Author*)

12.2 Marsh crocodile in Chotiari Reservoir (*Photo by Author*)

13.1 The *mohanna*s on Manchar Lake (*Photo by Author*)

14.1 Heavily polluted Keenjhar Lake (*Photo by Author*)

14.2 Noori's shrine in Keenjhar Lake (*Photo by Author*)

15.1 Keti Bunder's new mangrove plantation (*Photo by Author*)

15.2 Degraded mangroves in Keti Bunder (*Photo by WWF-Pakistan*)

15.3 Mangrove regeneration in the nurseries (*Photo by WWF-Pakistan*)

16.1 A baby green turtle making its way to the sea at Daran beach, Jiwani (*Photo by Author*)

it would be a one hour hike up the mountain to reach a ridge from where they could be spotted. I was not, however, informed about the steepness of the climb. Accompanied by two of the local tribesmen and Naeem Ashraf Raja, the wildlife and forestry specialist who then worked with Sardar Naseer Tareen, we climbed up a nearly dry stream bed and soon reached the cliffs. From there it was a near vertical climb to the top of these precipitous ridges. I climbed steadily, determined to reach the top, but my voluminous chador was giving me problems as it kept getting snagged on rocks and thorny branches. '*Baji*, please tie the chador before you end up killing yourself,' I was told by one of the tribesmen. I finally tied it around me and, mummy-like, reached the top of the cliff, to be rewarded with a spectacular view of the surrounding landscape.

With the wind blowing hard and the sun rising steadily, we hid behind some rocks and began scouring the landscape for any signs of the markhor. In less than ten minutes, my colleagues spotted a female markhor feeding on shrubs on a distant ridge and set up a tele-spotter for me to have a look. One can't get too close to these animals as they are extremely sensitive to smell and will scamper away at the slightest sense of human presence. As they camouflage well with their rocky surroundings (only the white marks near their hooves give them away), they are difficult to distinguish with the naked eye. The Afghan urial, on the other hand, is reddish in colour and is sometimes easier to spot when it is silhouetted against the blue sky. While the markhor prefer to remain among cliffs, the urial favour the plateaus above the cliffs and the terrain at their base.

The eminent biologist George Schaller wrote in his book, *Stones of Silence: Journeys in the Himalaya* (1980): 'Even on degraded land these species can survive, perhaps not at their most vigorous, but at least they can prevail until someday man has the wisdom to assure them a better future'.[2] Sardar Naseer has been fighting steadily since the 1980s to ensure that the Suleiman markhor and the Afghan urial have a safe future in Torghar. These large ungulates, along with leopards and black bears, once used to enjoy the free range of the mountains of northern Balochistan but by the mid-1980s, they were either exterminated or on the verge of extinction.

In 1984, Sardar Naseer contacted the United States Fish and Wildlife Service for technical assistance and they sent some of their experts to Quetta. Their discussions with Sardar Naseer led to the development of a plan to initiate a 'game guard' programme at Torghar called Torghar Conservation Project (TCP). The project was launched in 1985 with seven game guards who were former hunters who had agreed to put down their weapons. These men, selected from the local population, enforced a ban on hunting by protecting access to the area. Surveys were conducted and when animal populations had recovered sufficiently, a limited number of permits were sold to trophy hunters, primarily foreign hunters. The project continued to progress gradually over the years. Systematic trophy hunting took place every year from 1986 onwards, and the proceeds were used to hire more game guards and provide some financial assistance to the local community. At an initial figure of US$10,000 per markhor head and US$5,000 per urial head, the tribesmen began profiting from their conservation efforts. Older animals were carefully selected for culling in order to ensure that the herds' breeding rates were not affected. Additional game guards were hired every year; there are currently around 150 game guards protecting approximately 1,800 sq. km of Torghar. Today, the

Suleiman markhor's trophy fetches up to $60,000 and is one of the rarest in the world while the Afghan urial trophy is worth $16,000. Sardar Naseer Tareen told me that these big game hunters, mostly from wealthy European countries and the US, often fall in love with Torghar's rugged wilderness.

Around 20 per cent of the trophy fee goes to the government and the rest to the community. Sardar Naseer's pioneering efforts of using trophy hunting as a conservation tool is now being replicated by WWF-Pakistan in the north of Pakistan. The Torghar project is now recognized as one of the biggest success stories of conservation in Pakistan.

The TCP has directly resulted in a complete cessation of uncontrolled markhor and urial hunting in Torghar. There were an estimated 200 urial and under 100 markhor in the area when the TCP came into existence in 1985. According to the most recent survey conducted in 2011, the markhor and urial populations are now at around 3,518 and 1,181 animals respectively. Conducted by Michael R. Frisina, a professor of Range Sciences at Montana State University in Bozeman, and Tahir Rasheed, the national project manager SUSG in Quetta, the survey report states:

The Markhor population and its habitat are secure under the current management scenario. Maintenance of conservative hunting quotas and continued vigilance in minimizing overlapping habitat use between Markhor and domestic livestock are the keys to maintaining sustainability.[3]

Torghar is now home to the largest population of these unique animals in the world. Sardar Naseer's efforts to save the markhor have won him international acclaim: the Dutch have awarded him a knighthood in the Order of the Golden Ark and the French have awarded him their l'Ordre National du Merite. The International Council for Game and Wildlife Conservation (CIC) has also awarded the organization he founded with their Markhor Award for Outstanding Conservation Performance. Sardar Naseer Tareen is now considered to be an important figure in the field of conservation in Pakistan although, due to health issues, he now rarely leaves Quetta which is his current place of residence. He recently served as the chair of the IUCN's Sustainable Use Specialist Group (SUSG) and has also been a board member for WWF-Pakistan.

In April 1994, the Torghar Conservation Project was converted into an NGO called the Society for Torghar Environmental Protection (STEP) and registered in Balochistan. Largely self-sufficient since its inception, the project is the largest employer in the area. The NGO uses the money from trophy hunting to pay the salaries of its employees, for the community's medical care, for education (of local children who want to study in Quetta), and for development works such as the construction of small dams and installation of solar pumps. In 2015–16, STEP was allocated four Suleiman markhor and seven Afghan urial permits by Pakistan's Ministry of Climate Change. These were granted by the Secretary Wildlife of the Government of Balochistan. STEP utilized four Suleiman markhor and Afghan urial permits each for the hunting licenses issued to American and Turkish hunters.

The judicially managed trophy hunting has ensured the project's long-term survival, for the people's future is directly linked with the conservation of the wildlife of Torghar. As Sardar Naseer once told the local tribesmen: 'Torghar is like your mother, you live as long as she does ... therefore respect her water, her air, and her wild animals.' In his view, 'Projects like this only work when one gets on a first name terms with these mountain people. Over the years, I have had to deal with them on so many delicate issues that I know them thoroughly ... in the end it is their land and they can turn around and ask us to leave if they are not happy with us.'

Notes

1. M. H. Woodford, 'The Torghar Conservation Project', *Game and Wildlife Science*, vol. 21 (2004).
2. George B. Schaller, *Stones of Silence: Journeys in the Himalaya* (New York: Viking Press, 1980).
3. Michael R. Frisina and Tahir Rasheed, 'Straight Horned Markhor and Afghan Urial Population Monitoring on the Torghar Conservation Project, Balochistan, Pakistan' (Sept. 2011).

10
Salt Range: Rejuvenation of Soan Valley

I HAVE VISITED THE FAMOUS LAKES OF THE SOAN VALLEY, located atop the Salt Range and said to be the second largest deposits of mineral salt in the world, several times. This is a fragile and unique ecosystem created millions of years ago with the evaporation of the Tethys Sea. This arose when the Indian Plate collided with the Asian Plate, forming the subcontinent. The Salt Range extends from Mianwali in the west, to Jhelum in the east, and at its western extremes it forms a semi-loop creating the area called the Soan Valley. The average elevation of the valley is 2,000 ft above sea level.

Any traveller on the motorway from Lahore to Islamabad is sure to pass through the Salt Range. Unfortunately, due to the construction of the motorway which has provided easy access to this previously remote area, the government in recent years has permitted four cement plants to operate in the area. These water-intensive cement plants have not only destroyed the topography but are also posing severe health hazards to people of the area and depleting the groundwater. The local people would like the scenic valleys of the Salt Range to be converted into tourist resorts but the government is now allowing the construction of even more cement plants.

If you get off the motorway exit at Kalar Kahar and go past

the lake adjacent to the motorway, you will eventually find a series of three other lakes: Khabeki, Ucchali, and Jhalar which were declared a Ramsar Site in 1996 as globally important wetlands. The Ramsar Convention is an intergovernmental treaty that provides the framework for national action and international cooperation for the conservation and sustainable use of wetlands and their resources.[1]

The brackish/saltwater lakes in the Soan Valley are an important refuge for migratory birds from as far as Siberia. In winters, hundreds of pink flamingoes make Ucchali Lake their home, presenting a mesmerizing sight. I found the Soan Valley to be a very special place, unique in an arid country like Pakistan, with its abundance of streams, lakes, and hills. Soan is actually a Sanskrit word meaning beautiful. The valley itself, which is 76 km long and 19 km wide, is a peaceful and picturesque place steeped in history. It has been inhabited since ancient times; relics dating back to Hindu, Buddhist, Mughal, Sikh, and British periods have been found over here.[2] Indeed, visiting the valley is like stepping back in time with its rural hamlets and fragile old mountains. The valley actually functions as a basin: the rainwater runs down the sides of the mountains and collects in the lakes.

On my first few visits to these lakes, I learnt about the water problems facing the area as the lakes had begun drying up and groundwater tables were falling in the surrounding villages. Clearly the overuse of tube wells, which pump out water for irrigation and drinking purposes, had taken its toll on the ecosystem. Then came 2010's record-breaking rains and a minor miracle occurred in the Soan Valley. The lakes replenished themselves and the groundwater table rose. 'Even the wells that had been dry for around fifteen years have now been recharged,' a local villager told me excitedly on my last visit to the Soan Valley in 2011. There was no widespread flooding in this area, and the local villagers eagerly welcomed the heavy rainfall during the monsoon season of 2010.

Khabeki Lake had regained its original size (it had begun shrinking into a muddy puddle the last time I saw it). I hoped that the villagers would continue following the water conservation methods introduced by local NGOs and not resume overpumping the groundwater as they had earlier. WWF-Pakistan had initiated its conservation activities in the valley back in 2001 near Khabeki Lake where they established a field office. Later, the Pakistan Wetlands Programme focused its efforts on the three large lakes: Kalar Kahar, Khabeki, and Ucchali; the smaller Jhalar Lake; and the nearby Nammal Lake. This Salt Range Wetlands Complex, as it was termed by the programme, is also the core habitat for the endemic Punjab urial, a wild species of sheep.

The Khabeki wetland is brackish and muddy, and attracts many species of wintering birds, amongst them the endangered white-headed duck. During the last few years, the number of waterfowl visiting the lake has significantly decreased as the fragile biodiversity of the wetland has been seriously disrupted by hunting and fishing in the lake. Ucchali is a saltwater lake and is an ideal wetland for migrating greater flamingoes. Each year many migratory birds make a brief stopover of approximately three to four weeks to regain the strength to continue their long journey. The species of waterfowl that inhabit these wetlands in the winter months include the common pochard, mallard, purple heron, common coot, and little grebe. The common moorhen and purple swamp hen are permanent residents in the wetlands of Kalar Kahar.[3]

The other principal threat to this valley is deforestation. A long while ago the mountains surrounding Soan Valley were heavily forested. Indeed, one can still see the remnants of the thick forest in Sakesar, one of the highest peaks in the Salt Range, which is a protected area owned by the Pakistan Air Force (PAF) where no trespassers are permitted. This is where the PAF's main radar is located atop the peak which is 4,990 ft high. The mountains of the Salt Range were once covered with subtropical evergreen forests and tropical thorn forests but in recent years the villagers have cut down the forest cover on the surrounding slopes.

Deforestation has had a marked effect on the watershed areas and lowered the entire region's water levels. The Pakistan Wetlands Programme was monitoring the water table in the vicinity of the lakes to examine the sub-terrain water reserves and the relationship of the latter to the water level of the lakes. Thanks to the heavy rains of 2010, however, the lakes had been rejuvenated and the water tables in the neighbouring villages had also risen. Approximately 60 per cent of the population of Soan Valley is engaged in water-intensive farming and most of the land is under cultivation.

Over the years, the villagers have cut down the trees primarily for use as fuelwood because there is no provision of gas in this area. It is estimated that on an average 10–15 kg of wood is used daily during the summer by each household and this figure only increases during the winter season. Besides the heavy loss of vegetation cover, this also places a burden on the household economy as the daily cost of domestic fuel exceeds Rs 250 on average.

Forest fires have also destroyed many of the trees as there are mining activities in the area. Fortunately, the people of the Soan Valley are now learning about new ways of protecting their remaining forests. The site manager of the Pakistan Wetlands Programme, who was at the time implementing various conservation projects in the area, took me to visit a biogas digester that had recently been installed in one of the farms near Khabeki Lake.

Zafar Hayat owns a large farm in the picturesque hamlet of Dhok Khalan overlooking Khabeki Lake, and had been using a fiberglass biogas digester for a little over a year at the time I visited his home. 'Wood is very expensive in this area now,' he explained while showing me the biogas digester located near his house. 'And this is very cheap and easy to run. All we need is raw dung which is plentiful on the farm: all you need is a minimum of three large animals, either cows or buffaloes.' His wife, Bhag Pari, looked after the maintenance of the biogas digester which requires a regular feed of raw dung.

It is in her interest to do so because the biogas digester provides her with the gas she needs to cook for the household. 'I am free now from going into the jungle to look for wood. We have also saved on the wood my husband bought during his trips to the town. It gets very cold here in the winters and we would constantly need fuelwood. Now the gas burns all day and it is free!' she added excitedly. They had even installed a small heater in the house which runs on biogas. It was a spacious mud and stone home with several rooms all neatly maintained. Their kitchen was spotlessly clean without the usual black walls and smoky interior that comes with a wood-burning stove. From their verandah, one could see the green valley below. It was a relaxing, peaceful site for a home and I was tempted to sit on their charpoy and linger a while.

The family had contributed towards meeting 50 per cent of the cost of the biogas digester from their household savings. As it was a fibreglass digester, it was expensive and the total cost was around Rs 170,000. A brick biogas digester is half this cost but it is not as efficient as the fibreglass one although the technology is improving by the day. Zafar's neighbours were so impressed with this new technology that they had asked the Pakistan Wetlands Programme to help them to install biogas digesters on their farms as well. According to Zafar Hayat, 'I think if all the farms in this area were to install biogas digesters, then in two to three years the forest around here will regenerate. Biogas digesters are good and everyone should have them. Ours has really saved us from so much hassle.'

The Pakistan Wetlands Programme established two dozen biogas digesters in the area including the unique fibreglass plant at Zafar's home. Other community-based organizations, which have recognized the value of this intervention, have helped the local communities to install over a hundred more biogas plants in the area. These plants not only provide gas for domestic fuel but also supply them with a bio-fertilizer called 'slurry' which enhances soil fertility and supports plant growth and health. Slurry, with its many hidden benefits that support the entire ecosystem, is a good replacement for harmful chemical fertilizer.

The Pakistan Wetlands Programme also provided solar units to some selected households residing in the forested area. This not only helped to improve their lives in areas without electricity but also promoted the benefits of this alternate source of lighting. Now a majority of households in the Khabeki reserve forest area have installed solar units at their own expense.

Due to the high demand in the markets of Islamabad and Lahore for off-season vegetables, there is an increasing trend of cash crop cultivation in Soan Valley of cauliflower, potatoes, and onions—all vegetables which consume a lot of irrigation water. The Pakistan Wetlands Programme thought of encouraging the local farmers to plant fruit orchards instead to motivate them to move away from water-intensive crops. The programme planted fruit orchards in collaboration with a local NGO called the Soan Valley Development Programme which was founded by Gulbaz Afaqi around two decades ago.

Gulbaz Afaqi gave up a career in journalism in Lahore to return to his native village in the Soan Valley after a chance encounter with the renowned social scientist Dr Akhtar Hameed Khan. He inspired Gulbaz Afaqi to move back to his village and start a project to improve the lives of his fellow villagers through simple, indigenous solutions. In 1996, Gulbaz Afaqi founded the community-based Soan Valley Development Programme (SVDP). He decided that he could play a pioneering role in conserving the ancestral way of life of his people and help them find solutions to the problems they were facing as a remote farming community in the Salt Range. 'Big cities tend to eat you up. In rural areas you can achieve a sense of identity which is important,' he told me a few years ago. Gulbaz Afaqi conducted extensive water surveys with the help of experts and started an on-farm water management project by giving out small loans to farmers to construct proper water channels. His NGO also encouraged organic farming. He said of his return to his valley, 'I am much happier here than I ever was in Lahore. As a socialist who is interested in economic justice, doing this kind of work gives me an immense sense of personal satisfaction.'[4] Sadly, Gulbaz Afaqi is today quite unwell and has handed over the management of the SVDP to others in his organization.

In partnership with SVDP, efforts were made to conserve water through integrated watershed management and practising agriculture utilizing minimal quantities of water. Through the technical expertise of the Horticulture Research Station, Naushehra, a scheme was developed to promote fruit orchards requiring considerably smaller quantities of water. The climate of Soan Valley is considered to be supportive of fruit species such as olive, peach, pomegranate, plum, and apricot. Farmers were motivated and trained in orchard management followed by an exposure visit to Swat Valley to get a feel of the management of orchard produce. A number of progressive farmers adopted this initiative and developed orchards with financial assistance provided by the Pakistan Wetlands Programme and SVDP. Around fifty farmer families were supported in this intervention and now some of them are selling their produce on a commercial scale and receiving sizable returns.

An awareness campaign for schoolchildren was also conducted to highlight the ecological significance of the area in general and wetlands in particular. Informative booklets and banners were distributed while the celebration of world days linked to the environment and biodiversity were also organized. One of the outcomes of this awareness programme was a community-based watch and ward system to protect the forest and wildlife in the vicinity of Khabeki Lake. The local villagers began valuing the natural assets of the area and a visible decline was seen in the hunting of wildlife and waterbirds as well as in the felling of trees.

The Pakistan Wetlands Programme, in collaboration with its local partner SVDP, also focused on improving the quality of livestock which is the principal source of income for these communities. They organized training camps, held at Khabeki and Ucchali lakes, to educate women about livestock management practices. The deworming of livestock in collaboration with the National Rural Support Programme was carried out as well to promote the health of livestock in the valley. Secondly, they also wished to avoid possible disease transmission from domestic to wild ungulates like the Punjab urial which is endemic to this area. WWF's Punjab urial conservation initiative was also established with the deployment of community wildlife guards to protect vulnerable resources. Unfortunately, I was not able to spot a Punjab urial during any of my visits as it has become so rare and its terrain is so vast.

The existing WWF-Pakistan field office at Khabeki Lake was supposed to be upgraded to a Wetlands Information Centre with a birdwatch tower, library, conference hall, and ecological exhibits. However, with the phasing out of the Pakistan Wetlands Programme, the establishment of the Khabbeki Conservation and Information Centre could not be completed on time. The programme had taken up some other initiatives to promote ecotourism in Soan Valley and several lake-view points, bird hides, and campsites were set up at some of the lakes.

This projection of the area as a tourist site by WWF-Pakistan has now become one of the priorities of the Punjab government. Khabeki and Ucchali lakes are being developed as tourist sites to promote tourism in the Soan Valley. I once took a walk around Ucchali Lake which is the largest saltwater lake in the valley; the soft earth around the lake had streaks of white salt on it and the water in the lake was a strange greenish-yellow. The local people believe that there is a volcano hidden beneath the surface of the Ucchali Lake which causes the colour of the water to constantly change. The government has now constructed a jetty for boating in the lake and developed different tracks for walking around it.

Unfortunately, uncontrolled tourism, infrastructure development, and exhaustive agricultural activities are taking their toll on the natural treasures of Soan Valley. The Tourism

Development Corporation of Pakistan needs to promote eco-friendly tourism in the area otherwise the valley may face irreversible ecological damage. Farmers also need to put an end to the excessive use of fertilizers and pesticides as well as groundwater extraction which is causing further environmental degradation. More needs to be done to safeguard the Soan Valley. As Gulbaz Afaqi once put it, 'My vision for the Soan Valley is of a place where farming is done in a special way that conserves water and soil, the traditional way of life, and the wildlife.'

Notes

1. 'Khabeki Conservation and Information Center (KCIC)', *Wetland Link International*. Website: <http://wli.wwt.org.uk/2016/03/members/asia/asia-members/khabeki-conservation-and-information-center-kcic/>
2. 'Soan Sakasar Valley', *Revolvy*. Website: <https://www.revolvy.com/topic/Soan%20Sakaser%20Valley&item_type=topic>
3. 'Salt Range Wetlands Complex', *Pakistan Wetlands Programme*. Website: <http://pakistanwetlands.org/srwc.php>
4. *Green Pioneers Stories from the Grassroots* (Karachi: UNDP-Pakistan and City Press, 2002).

11

Indus: Saving the Blind Dolphins

I ARRIVED IN SUKKUR TO FIND A DEAD INDUS DOLPHIN lying in the garden of the Sindh Wildlife Department's office. Uzma Khan, a biologist who works for WWF-Pakistan, was preparing to conduct a post-mortem to ascertain the exact cause of death. Quite traumatized, I knelt above the dolphin, its delicate silver-grey skin with a hint of pink shimmering in the afternoon sun. I was surprised to see how fragile and small it really was at close proximity. 'It's probably an infant, no more than three or four years old,' said Uzma (Indus dolphins can live up to thirty years). It had died the previous day in the course of a rescue attempt. During the autopsy, Uzma discovered that its rostrum had snapped, perhaps through entanglement in the fishing net during the rescue, a common cause of death for these dolphins.

During the monsoon season, the Indus dolphins routinely swim into the many connecting canals of the Indus when the gates of the barrage are opened to maintain the water level of the river. When the canals are closed in winter for de-silting and the water level falls, the dolphins are stranded and frequently die of starvation or are inadvertently drowned when stuck in the nets of the local fishermen who fish extensively in the closed canals. The Sindh Wildlife Department in collaboration with WWF-Pakistan has set up an Indus Dolphin

Rescue Unit to come to the aid of the trapped dolphins and physically transport them back to the Indus.

I had come to Sukkur in 2009 to write a story about the dolphins because for some inexplicable reasons over 50 dolphins had become trapped in the canals near Sukkur that year. Was it a mass migration? Were they starving due to a paucity of fish in the river? Or was there just not enough water in the Indus that year? None of us knew the answers, including François-Xavier Pelletier, a film-maker from France who has devoted his life to filming river dolphins in the Ganges and the Amazon rivers. He was also in Sukkur at the time, planning a detailed documentary on the Indus dolphins.

This was not François' first trip to Pakistan; during his initial visit almost ten years ago to make a film on the *mohanna*s (or fishermen who live on the river in their houseboats), the bus he had taken in interior Sindh was attacked by dacoits. 'They killed the driver, raped a woman on the bus, and robbed all the passengers except me,' he told me. 'They said to me: "Welcome to Pakistan".' 'It's a crazy country,' was all I could mumble. François has already written a book on the 'sacred dolphins' based on his travels in India, Nepal, Bangladesh, and Pakistan. He told me that a very similar species of blind dolphin could be found in the Ganges and Brahmaputra rivers across the border. 'What is really sad, however, is that because of all the lawlessness in Sindh, the boat people are abandoning their traditional way of life. For centuries they lived on this river, now they have to search for alternative livelihoods,' he said about the *mohanna*s.

Before the construction of the barrages, both the fishermen and the Indus dolphins enjoyed free-range of the mighty Indus River. A cousin of the sea dolphin, the Indus dolphin is thought to have come into existence millions of years ago, when the Indus River was cut off from the large Tethys Sea as a result of the collision of the Central Asian and Indian subcontinent's tectonic plates. The sea-dwelling dolphin was trapped in a river system that flowed out on to a new land mass. The marine dolphin was thus obliged to become an entirely freshwater species, and centuries of living in the turbid waters of the Indus made its eyes redundant. This is the theory of how the Indus dolphin, one of the world's four remaining freshwater cetacean species, came to inhabit the Indus River (the others are found in the Amazon, Ganges, and Yangtze rivers). According to genetic research, the Indus dolphin is essentially a 'living fossil' that has not shared a common ancestor with any other living creature for around 25 million years! It has, therefore, the most unique cetacean lineage in existence.[1]

In nineteenth century British India, the ecology of the Indus River began to change with the construction of barrages and dams. Today, there are seven barrages and several dams on the river that make it one of the most formidable irrigation systems in the world. The virtually impenetrable barrages have carved up the Indus dolphin's home range. The current remaining schools of the Indus dolphin are found largely in the Sindh area, between the Guddu and Sukkur barrages. This 170 km stretch of the Indus River was declared an Indus Dolphin Reserve by the Government of Sindh in 1974, and it has also been proposed that it be declared a World Heritage Site. There are today an estimated 857 dolphins in the reserve (last recorded in 2011).[2]

Every five years, WWF-Pakistan along with the Sindh, Punjab, and KPK Wildlife Departments and other NGOs, conduct an Indus dolphin population survey of the river. According to the findings of the latest survey, conducted in 2011 after the massive floods of 2010 (with one section covered in 2012 due to security concerns), a rise in the dolphin population was observed in some of the river sections, indicating that its distribution had increased (previously it was concentrated in the river section between the Sukkur and Guddu barrages).

According to Uzma Noureen, the project coordinator of the Indus River Dolphin Conservation Project:

> In the river section between the Taunsa and Guddu barrages, we recorded 465 dolphins. This section previously has a record of 259 dolphins according to the 2001 survey. In the river section between the Sukkur and Kotri barrages, we recorded 34 dolphins whereas in 2006 only 4 dolphins were seen.

This is a direct count: a boat goes out on the river and surveys the dolphins, spotting them when they surface for air. During the last survey completed in 2006, the estimated population of Indus dolphins was found to be around 1,600 and now it is around 1,452 which Uzma Noureen says is 'a slight decrease'. This decrease in dolphin population could, however, be due to the higher water level in the Indus after the floods, providing the dolphin population an opportunity to disperse. She went on to explain that after the floods of 2010, many channels had been reactivated and they could have been missed during the survey:

> We survey during low flow season, when the possibility of spotting dolphins is high because the dolphin population is concentrated in the main river stream, but after the floods of 2010, there was more water in the river and the river's extent had increased, dispersing the dolphin population and

increasing its distribution. If after the floods we found slightly fewer dolphins overall, we can't really say that the population has decreased.

In Sukkur, we made our way to the Indus Dolphin Reserve to do some filming of the river and of the dolphins as they surfaced for air. We crossed the metal Lansdowne Bridge, built by the British as a marvel of engineering at the dawn of the twentieth century, over the Indus and into Rohri (opposite the town of Sukkur). I had always wanted to visit the mysterious shrine of the seven virgins, built on a limestone mound overlooking the river, and finally got an opportunity to do so that day. There is really something very peaceful and captivating about this stretch of the Indus as it flows through Sukkur. We parked the car near the river and walked through a grove of old date palms. The place resonated with history: ancient tiles of glazed blue and white bricks (fallen off the shrine above) were littered on the ground, a huge neem tree stretched its shady branches at the entrance, and above us was the shrine with its many gravestones which had stood for centuries as sentinels.

We climbed up the many limestone steps to enter through a musty doorway with glazed tilework decorating its archways and up a narrow, dark corridor and into the sunshine again on the roof terrace. The caretaker of the shrine narrated:

This shrine is 1,300 years old dating back to the time of Raja Dahir, a cruel ruler who had attempted to kidnap these seven virgin sisters and take them to his fort across the river on the other side. They prayed to God to save them and the rock opened up and they were buried alive and their honour was saved.

The myth of the seven virgins seems to be a popular one in Islamic folklore. At any rate, there are no actual graves of the seven virgins, although rumour has it that there is a secret room somewhere below the shrine where their graves are to be found. The many graves on top of the shrine are those of the notables of that era, some of which date back to the early Muslim invasions of Sindh. I was intrigued to see some of the more elaborately carved tombstones adorned with verses from the Holy Quran along with the symbol of the swastika (originally a Hindu motif).

Below us was the Indus Dolphin Reserve, stretching for miles in either direction. We climbed back down and hired a nearby wooden boat (there are no motor boats in the area) and asked the oarsman to take us to the middle of the river. Before long we spotted the dolphins surfacing for air. They leap out of the water in a graceful arc but you have to be very alert because it only takes them a couple of seconds. There were dozens of them. This was truly their home territory, in such close proximity to the two towns of Sukkur and Rohri that had crept up along the banks of the river. We neared the island of Sadh Bela which is now a Hindu shrine. Two hundred years ago, a Sadhu had come there to meditate and pray, and when he died a shrine was built on the deserted island in the middle of the Indus. Every June a large festival is held in his memory which is attended by thousands of Hindu and Sikh pilgrims. So there you have it: a Muslim shrine on one shore and a Hindu shrine on the other, both peacefully coexisting with the dolphins in between.

By now the sun was setting and I leant back comfortably on a *rilli* (traditional handmade blanket) that had been placed on the wooden boat to watch the sky turn orange while the river became gold. In the winter, the Indus is a calm, vast river stretching for miles and the boat swayed gently in the mild current. It was magical: the entire river shimmering for miles into the distance where a small yellow fireball was disappearing below its surface. The sun finally set but then the sky became even brighter because its rays were reflected off the clouds and bounced back down on to the surface of the gilded river. I felt I was part of it all, and yet realized how little I mattered here on this ancient river that has seen so many civilizations rise and fall on its banks. Would the Indus dolphin that has survived for millennia in this river, have a future?

The following day I was sufficiently fortunate to take part in two rescues not far from the Indus River in Sukkur (sometimes the dolphins swim far into the canals and it takes almost an hour by road to bring them back to the river on a stretcher). Hussain Bux Bhagat, then deputy conservator of the Sindh Wildlife Department, was preoccupied that day arranging for visitors who were flying in to Sukkur to see the official rescues. Consequently, Uzma Khan and I were given a freehand to take charge of these two rescues. Two dolphins had been trapped in a small tributary off the Indus, a couple of miles from the Indus Dolphin Reserve near Lansdowne Bridge. We found them swimming together in a large pool of water just below the road. Uzma decided that it would be far better not to transport them on a pickup (used for most rescues) but to carry them back to the river on foot in a stretcher. 'It's the noise that really bothers them. In a car, being jostled around and the sound of the engine [is] very traumatic for them. They are mammals and, therefore, keeping them out of water for a long period is not so much of a problem,' she explained.

She outlined the routine to the fishermen and wildlife department officials who had gathered in a circle:

Do not carry them holding their flippers and beak; those are very sensitive areas. Only one person will dive in and pick them out of the water. Ensure their blowhole remains clear; do not put water around that area. Don't talk or make too much noise when we are carrying them to the river. Above all, remain calm!

Two stretchers (with foam mattresses and wet towels) were readied and the fishermen began laying a net to try and get the dolphins to swim to the shallower parts of the pool. In an earlier rescue I had been unnerved by the large number of people who had gathered around and obstructed the stretcher so I instructed a few of the guards accompanying us to ensure that no one climbed down the steps that ran down from the road. The cameramen from Geo TV were most amused by my bossy approach.

The net closed in on the dolphins and there was a small cry as one of the dolphins headed towards it (we have to ensure they don't get entangled in the net). The dolphin, however, seemed to have jumped over the net and rapidly swam away. Uzma told the fishermen to go for the second dolphin and attempt to capture the other one later. Nazir, the veteran dolphin catcher who worked for the Sindh Wildlife Department, suddenly jumped into the water and caught the dolphin in his arms. Holding the dolphin gently, following the detailed instructions from Uzma on the shore, he walked slowly towards us. The dolphin was laid on to the stretcher, Uzma covered it with a wet towel and, instructing one of the fishermen to pick up a bucket, we began the long walk to the river over fields of vegetables grown by local farmers. The dolphin seemed to remain calm. Richard Garstang, then conservation advisor at WWF, had explained to me that once out of the water, the dolphin 'experiences a blackout'. I hoped so, for its sake: imagine living in a river all your life and then suddenly being hauled out of it.

The Geo cameramen, who were both from Sukkur, had told us earlier how they used to swim in the river when they were younger and that the dolphins would play alongside with them. 'They were very friendly and completely harmless,' they recalled. They too were fascinated to see them taken out of the water. In less than no time, we arrived on the banks of the river. Uzma took out the measuring tape and quickly noted the dolphin's size. The stretcher was then taken into the water and Uzma asked Nazir to just let it swim away without touching it. We waited for it to surface but it took its time until, eventually, Nazir spotted it several metres away.

Uzma later told me that it was a female dolphin. Coincidentally, the other one turned out to be a male. I thought it very romantic that they had been stranded together and now would be rescued together. Although I had hoped the female would wait for the male but since they can use sonar to communicate with one another, I was certain that they would find each other sooner rather than later. It seemed like the male dolphin had returned to search for the female and had been caught by the time we arrived back at the pool. He was immediately laid out on the stretcher and we began our long walk back to the river. He too was successfully set free and we stood watching for him for a long while. In the muddy waters of the Indus it is very difficult to tell the dolphins apart.

It had been a tiring but satisfying day and I left Sukkur the next morning with happy memories. The Indus Dolphin Rescue Unit has to date rescued a total of 117 dolphins. They are performing an incredible service by helping to rescue these dolphins, precious 'living fossils', and among the most endangered mammals on earth.

Notes

1. GillianT. Braulik, 'Conservation Ecology and Phylogenetics of the Indus River Dolphin (*Platanista gangetica minor*)', PhD thesis (University of St Andrews, 2012).
2. 'Indus River Dolphin', *WWF-Pakistan*. Website: <http://www.wwfpak. org/species/indus_dolphin.php>

12

Desert Lakes: Chotiari Wetlands

THE CHOTIARI WETLANDS COMPLEX, AS IT IS KNOWN, IS ONE of the most biologically diverse places I have ever visited in Pakistan and is home to marsh crocodiles and hog deer. Located at the junction of deep lakes and shallow marshes; and bordered by sand dunes, agricultural land, and a riverine forest; it is situated in the Achro Thar Desert around 30 km from the town of Sanghar in interior Sindh. The Chotiari reservoir was actually constructed out of a natural depression along the left bank of the Nara Canal in the 1990s. The idea was to store rainwater and floodwater from the nearby Indus River to use in times of drought. The reservoir area thus comprises a number of freshwater and brackish lakes which provide fish for the local fishing communities and are a refuge for wildlife. As the lakes are a feeding and nesting ground for a variety of birds, this is a well-known hunting ground for shikaris from all over Pakistan.

The local people who live here depend on fishing, agriculture, and livestock for their livelihood. Many are followers of Pir Pagaro, the spiritual and political leader who lived in nearby Khairpur district and passed away in 2012. Every few years, he would visit the area and be accorded a warm and festive welcome. Around thirty villages are located around and within the reservoir. Most of the local people live below the poverty line and the area lacks basic social services and infrastructural facilities. The people live in cone-shaped thatched huts without any electricity or piped water and are entirely dependent on natural resources for survival.

When I arrived in Sanghar, I took the metalled road leading to the Chotiari Wetlands Complex. The lakes, which shimmered in the hot sun (I had gone in the late summer months), were surrounded by the sandy-white Achro Thar Desert. There were huge sand dunes that extended from north to south. In between was a mosaic of interlinked fresh and saline water ponds and bays that occupy an area of about 18,000 ha. When flooded, the wetlands have a water storage capacity of about 0.75 MAF and, in the process, inundate an area of approximately 160 km².

Seven *deh*s or village clusters are included within the boundaries of the Chotiari Wetlands Complex. The population is primarily distributed around the margins of the lakes Makhi, Haranthari, Bakar, Akanvari, Khadvari, and Phuleli. The population of the area is around 14,000 people who are mainly Sindhi speaking. The livelihood of the local community depends upon commercial fishing with small-scale agriculture, livestock rearing, and other activities such as the making of mats using reeds, harvesting herbs and medicinal plants, and fuelwood collection. The fertile irrigated lands on the periphery of the complex are used for growing cotton, wheat, and fodder crops.[1]

The local people were not happy with the construction of the Chotiari Reservoir. Some had still not received any compensation despite having had lost their land when submerged by the reservoir. They told me that the construction of the reservoir, with its long embankments and dykes built to convert the various lakes into a single large reservoir, had disturbed the natural flow of water into the lakes and spoiled the water quality of some of them.

The reservoir was also supposed to increase the production of fish in the area but that has not occurred as there is an uneven supply of water into the reservoir. When the water level falls, the fish production too declines. Today, there is a decreased fish catch due to the loss of flora in the reservoir which has had a significant impact on the livelihoods of the fishing community. When I visited the reservoir in 2009, the water level was so low that the people living there were demanding that the government should release more water into the reservoir. 'Please tell the outside world about our plight,' they implored me.

The construction of the reservoir has also disturbed the habitat of the unique wildlife species found in the area. The smooth-coated otter can now only be found in small pockets and the hog deer is also an endangered species in the area. Degradation and shrinkage of habitat is one of the principal reasons for the decline in the hog deer population. A total of 80 avian species have been sighted in and around the wetlands complex. Species requiring special protection include the marbled teal, sind babbler, fishing eagle, cinnamon bittern, glossy ibis, purple heron, grey and black partridges, and the common buzzard whose population has decreased and is rated as scarce.[2] The biological diversity of this area had drawn the attention of WWF-Pakistan who was working there under their Indus for All Programme.

The fishing communities living around the reservoir, represented by the Pakistan Fisherfolk Forum NGO, had become partners with the Indus for All Programme which was trying to introduce conservation activities and improve livelihoods in the area. The programme helped organize the local communities into nine community-based organizations (CBOs). The CBOs were sensitized about the use of natural resources and given a training course in management. The programme had also installed solar energy units in selected mosques and opened up a vocational centre in Chotiari where girls were trained in sewing/cutting and hand embroidery work; they had learnt to make embroidered bags, wall hangings, and cushions. Workshops on health and hygiene for women were also conducted.

In order to sustainably manage the precious biodiversity of the Chotiari Wetlands Complex, WWF-Pakistan later completed the Chotiari Conservation and Information Centre. Built in traditional style with mud walls and thatched roofs, the centre is now up and running. The Sindh government even provided support for livelihood improvement in the area and is supporting the centre. The spacious centre has a biodiversity resource space, a fisheries resource room, a conference hall, several educational environmental exhibits, and a community handicrafts outlet. There is also a cafeteria and an administrative office.

Surrounding the centre is a botanical garden and wild-plant nursery and a birdwatching tower. Energy is generated through a solar electrification unit at the site. The water supply comes through a centralized drip and rainwater harvesting system. The centre provides a venue for the local communities to meet and hold discussions on conservation and seeks to raise environmental awareness and promote ecotourism. It is performing extremely well and has hosted many visitors, which in turn has generated income.

Nature clubs were also formed at the local schools to spread awareness amongst children. Over thirty schools in and around Sanghar were involved in nature walks, tree plantation activities, competitions, debates, and tableaux about the environment. I visited one nature club in the area, located in the Government High School outside the town of Sanghar. The school was remarkably clean and there were garbage bins bearing WWF logos all over the campus. The grounds were also very green, with small shrubs and trees. I was quite impressed to see all the greenery. 'The children, especially the members of the Nature Club, keep the school very clean,' said Khan Mohammad, one of the teachers at the school. The classroom where the Bird Lovers' Model Nature Club was located was adorned with colourful paintings of animals and birds.

I noted the environmental education kits stored in the cupboards for use during the Nature Club meetings. Outside the classroom was a Bird Lovers' corner where the children could feed birds. 'We take the students to Chotiari Reservoir and teach them about fish, birds, and animals,' said the schoolteacher who had been trained in environmental education by the Indus for All Programme.

Two of the teachers at the school were master trainers and in a position to impart environmental education training to the other teachers. The Model Nature Club was quietly bringing about change in Sanghar as the children, who learned about hygiene and environmental issues, would go home and teach their parents about the same thus spreading the message of conservation.

The sun was setting when I went to visit a village located across the nearby Nara Canal, a few kilometres from Sanghar town. The village, called Ghulam Hussain Laghari, is right outside Makhi forest (named after the small bees which make their honeycombs in the trees inside the forest). Makhi forest was once famous for its rich reserves of quality honey, commercially valued wood, and plants of medicinal value. The forest was also the stronghold of the freedom movement launched by the Hurs (followers of Pir Pagaro) against the colonial British power during the 1930s. During the uprising, the Hurs would hide in the forest. To suppress the Hur Revolt, the British rulers converted a large part of these woodlands into agricultural areas.[3] The rise of the water level in the nearby Chotiari Reservoir had also damaged much of the forest.

To reach the village, we had to clamber on to a small wooden boat which was then pulled across the Nara Canal by a boatman holding onto a wire stretched across the two banks. I had never seen such a contraption before but it seemed to work quite effectively for we reached the other side in no time at all. Members of the Makhi Development Organization (a community-based organization registered in 2008) were waiting for me on the other side. The sun was beginning to set when we crossed the canal and the calm water was shimmering whilst a thick forest beckoned me from the opposite side. I felt as if I was about to enter into some enchanted forest teaming with wildlife.

The local people living in and around Makhi forest are completely dependent on the forest for their livelihood. They settled in the area when they moved here in the 1970s. The CBO's President, Khalid Ali, said:

Over the years, we killed almost everything in the forest—all the animals and birds, and we chopped down so many of the trees. Then the WWF people arrived here a few years ago and they started talking to us, explaining that the forest belongs to us and we have to save it for our future well-being. ... The Lagharis are in fact traditionally famous as hunters and now we are saving the animals!

Hog deer, desert hare, jackals, otters, wild boar, and crocodiles were once found here in abundance. 'We are very aware of the fact that we need to save these animals now. We don't want to just read about them in story books; we want to see them alive in the wild,' said another member of the CBO. There were half a dozen villages involved in the Makhi Development Organization (the size of the villages range from 1,000 people to 100).

The CBO was planting more trees in the forest with the help of the Indus for All Programme. They were also introducing biogas units, which run on buffalo and cow dung, in the largest village. Two units, which could provide gas for the stoves and run energy-saver bulbs for three houses each, were being installed when I visited. There was an emphasis on the use of biogas as it can greatly help decrease the amount of wood cut for fuel purposes.

Makhi forest is located in a very remote area with no hospitals or schools nearby (the nearest boys' school is 2 km away). However, the people feel they are rich in resources: the honey from the forest is highly valued and at that time they could sell it for as much as Rs 700 per bottle. 'We get demands from all over the country for this honey,' one villager pointed out. The forest is also used for grazing by the villagers' livestock; goat droppings actually help regrow the trees by spreading seeds. 'It's the people who are the biggest threat to the forest. They cut the wood and sell it in nearby towns and cities. But we've managed to put a stop to that. It's under our control and we have stopped the logging,' said Khalid Ali. The forest is gradually regenerating and the animals have begun returning.

Chotiari has great potential for ecotourism given its unique biological diversity and rich cultural heritage. For its part,

WWF-Pakistan has undertaken invaluable work in the area to spread awareness. I left Chotiari with happy memories of its sand dunes, clear and brackish lakes, riverine forest, and especially its environmentally conscious people. I fondly hoped that it would one day become a well-known tourist destination in Pakistan for those seeking adventure in the wilderness.

Notes

1. 'Chotiari Wetlands Complex', *foreverindus.org*. Website: <http://foreverindus.org/ifap_sites_chotiari.php>
2. *Detailed Ecological Assessment of Fauna, Including Limnology Studies at Chotiari Reservoir* (WWF-Pakistan/Forever Indus, 2008).
3. 'History of the Hur Movement', *Jadamsindhu* (Feb. 2011). Website: <https://jadamsindhu.wordpress.com/2011/02/05/history-of-the-hur-movement/>

13

Toxic Waters: Manchar Lake

I ARRIVED AS THE SUN WAS BEGINNING TO SET WITH Manchar Lake presenting a picturesque sight, shimmering in the fading light. A few *mohanna*s (as the local fishermen are called) were straining at the bamboo poles on their wooden boats, pushing them soundlessly and determinedly into the water to propel them on their way home. The lake stretches for miles in every direction and the shoreline at the other end is barely visible; it is more like a mini-sea than a lake spread over 200 sq. km. During the monsoon rains, it can expand to as much as 500 sq. km and was once considered Pakistan's greatest water treasure.

Manchar Lake is Asia's second largest lake and Pakistan's largest shallow water natural lake. Located at a distance of about 18 km from Sehwan Sharif town in district Dadu in Sindh, it is a vast natural depression surrounded by the Kirthar Hills in the west, the Laki Hills in the south, and Indus River in the east. The lake is only around 3 m deep and it is difficult to believe that beneath its smooth and silvery surface, the lake water is full of toxins which have caused migratory birds to flee, destroyed the local agriculture, devastated the *mohanna* community, and brought death to people living in nearby cities.

In the summer of 2004, hundreds of people in Hyderabad city in Sindh (and further down the Indus River) fell ill after

112

drinking the poisonous water of Manchar Lake which had been allowed to enter the city's water supply. In Hyderabad alone, around forty people, mostly children, died from severe diarrhoea caused by gastroenteritis. The lake's water reached the city via the Indus River, then in spate thanks to the heavy rains that had finally broken an extended period of drought in Sindh. A fact-finding report conducted by Pakistan's Human Rights Commission squarely blamed the lake's toxic waters for the disaster.

When I visited the lake in 2007, it was again full of water, having rained heavily the past year. The lake is largely fed by hill torrents so the rainwater had rushed down and filled up the lake. This happened again during the heavy floods in Pakistan in 2010 when the lake's toxins were briefly flushed out by the massive flooding in the area which recharged it. According to a report on Manchar Lake prepared by the environmentalist Naseer Memon:

> Inflow [in]to the lake is very much erratic and unreliable since it depends on flood flows. Mean annual rainfall in this area is only 4.43 inches (112.5 mm) against evaporation of about 80 inches (2,000 mm). Therefore, very little run-off is generated within the catchments of hill torrents during dry or average years. Hence the lake is recharged by this source only during wet years.[1]

The other principal source of freshwater for the lake is the Indus but the river itself is facing water shortages.

When I visited Manchar Lake, there were sizeable quantities of fish for sale in the small marketplace, located in the settlement on the banks of the lake, as we left the narrow strip of asphalt that veers off the main road from Sehwan Sharif.

This settlement comprised largely of mud houses and a main bazaar and was reeking of rotting fish and open sewers. The market was crowded with the local people sitting in groups sipping tea, watching television, and discussing local politics. I noted a number of NGOs with signs up that announced that they were working for livelihoods in the area. I was, however, more interested in meeting the *mohanna*s, the traditional fisherfolk who still live in their wooden houseboats, as they have for centuries, in harmony with their natural environment.

The only difference now is that their environment has become heavily polluted. Today there are only around four dozen floating homes left on the lake. Back in the 1980s when the 'water was sweet', there were approximately 20,000 people living on the lake in over 400 boats. 'The annual fish catch on the lake has dropped dramatically from 2,300 metric tonnes in 1944 to 700 in the 1980s,' said Mustafa Mirani, vice-chairman of the Pakistan Fisherfolk Forum, an NGO working to promote the rights of the local fishing community. 'Today, it is no more than 75 metric tonnes,' he estimated.[2]

'Most families moved because they were starving here. The water is spoilt—the natural vegetation in the lake has gone, the fish have died mostly. There is nothing here for us now. We would even leave tomorrow if we could but we have no money at all,' said Nazeer, a *mohanna* who lived with close relatives in a cluster of eight boats. These large houseboats with their intricate handcarving serve as mobile homes for the *mohanna* community. A peek inside one revealed bedding, a makeshift kitchen, shelf space for storing food and household items, and even a cradle for the baby. Although the *mohanna*s don't own any land, they do own the boats which have been passed on from one generation to another.

There are a number of floating villages occupied by different communities on the lake.

According to Naseer, the water in the lake began becoming polluted in 2000, probably following Sindh province's six-year long drought. Manchar Lake was once a tourist destination; a wetland which was a haven for migratory waterfowl and home to a thriving community of *mohanna*s and agriculturalists who raised crops around the lake's shallow bed. Over the years, a paucity of adequate rainfall and an influx of chemical effluents released into the lake began playing havoc with it.

Manchar is seriously affected by the construction and enlargement of artificial channels linking the Indus River with the lake and the construction of flood embankments to the north. 'The Main Nara Valley (MNV) Drain (built during the construction of the Sukkur Barrage) brings a considerable amount of saline water into the lake,' Naseer Memon points out. The fishermen recall that their forefathers had once enjoyed fresh, clean water in the lake that enabled them to make a good living from the ample fish stock available.

To make matters worse, the controversial and poorly designed Right Bank Outfall Drain (RBOD, as the remodelling of the Main Nara Valley Drain is now called) began dumping industrial effluents (from factories up north) and agricultural run-off into the lake. The authorities assumed that freshwater from the Indus and from the hill torrents during the rainy season would dilute the effluents. That was a miscalculation because Manchar's two sources of water do not provide it with sufficient water to wash out the effluents on a regular basis. Besides, flows from the Indus are drying up as a consequence of all the barrages and dams located upstream, while rainfall in Sindh in recent years has become extremely erratic due to climate change.

The misconceived project destroyed Manchar Lake and its fish and has brought untold misery to the indigenous fishermen community. According to water experts, people who drink Manchar Lake's water can contract waterborne diseases, and in the long run develop serious illnesses. Environmentalists want the present drainage from the Right Bank Outfall Drain into Manchar Lake to be immediately halted. The government, having realized the damage caused by the drain, decided to divert it into the Arabian Sea. This project, known as the RBOD-II, has witnessed lengthy delays and cost overruns. It has been suspended for years due to lack of funds thus contaminated water continues to flow into the lake untreated and unabated.

What was also alarming to hear was that almost 40,000 ha of land in Sindh is currently being irrigated by the waters of Manchar Lake. Imagine the consequences of all the crops and vegetables around the lake absorbing the toxic chemicals in the water from effluents discharged from factories and agricultural run-off. These crops and vegetables are probably harvested and sold in towns and cities throughout the province. Hard chemicals from the effluents enter the food chain and are known to be carcinogenic.

Naseer Memon says Manchar Lake is nature's gift to Pakistan and, in order to save the lake, the following actions need to be taken. A comprehensive revised plan for rehabilitation and conservation of the lake should be prepared with the active participation of the local communities and experts. The inflow of saline effluents from MNV Drain should be delinked from Manchar and the freshwater flow from Indus should be enhanced and regularized. Water quality monitoring should be undertaken with greater frequency and the results should be made public. Health facilities should also be provided to save human lives.

The *mohanna*s themselves are not taking any chances with the polluted lake water. At the time of my visit, they bought their drinking water from the bazaar and stored it on their houseboats in large plastic containers. The livestock in the area, however, drank from the lake and was infected with all manner of diseases. I later found out that the government had installed over twenty water filtration plants in different locations near the lake but many, due to lack of maintenance, soon became dysfunctional.[3]

I visited some more of the houseboats, many of which were inhabited by women, and noticed that there were hardly any men around. 'Most of the young men have gone to the coast to find work in the fishing trade. Only the women, children, and old people are left here now. It is the same situation all over the lake. Those who can, they migrate,' said the old grandmother while her daughter lit a fire on their houseboat to cook the evening meal. 'We have lived on this lake for almost seven generations but now I don't see much of a future for us here.' In the past decade or so, many of the fishermen have moved to Gwadar and Pasni in Balochistan to fish in the sea, which is much more difficult because it means venturing into a deep-water environment.

It was by now getting dark so we made our way to the shore. We saw a man preparing to slaughter a duck he had caught on the lake that day. In the past, the lake was home to numerous migratory species of bird. It being the first wetland on their route, Manchar Lake was a safe wintering habitat for thousands of migratory waterfowl. Wildlife experts say that the pollution in the lake has caused a dramatic fall in their numbers. Migratory birds continue to visit the lake but fly away after an overnight stop.

As we made our way back to the main road where our car was parked, we noticed a number of small children carrying wood. The *mohanna*s had told us that selling fuelwood in the bazaar was now one of their principal sources of income. One little girl was bent over, carrying a large bundle of wood on her head, her clothes no more than rags. There were obviously no schools in the area. Later I learnt that a boat school was established for the fishing community at Manchar Lake. Each morning a boat would pick up students and they would be taught within the boat for five hours during the day.

At the time of my visit, the local communities had other, more immediate problems to cope with: the prevalence of tuberculosis, anemia, malnutrition, skin disease, gastroenteritis, and waterborne disease was widely reported. These people are amongst the poorest of the poor in this country with no voice whatsoever, and the toxic water of Manchar Lake is slowly but surely killing them off. The Sindh government did not seem to be interested in solving their problems until, miraculously, the former Chief Justice of the Supreme Court in Pakistan, Iftikhar Muhammad Chaudhry, accepted a suo motu notice regarding the contamination of Manchar Lake before he retired in 2013.[4]

The Supreme Court is currently pursuing the case; the judges have observed that the disposal of polluted water into the lake from the Right Bank Outfall Drain is not only destroying the beauty of the lake but also depriving the local fishermen of their livelihood. They noted that the toxic water is affecting the marine life and causing an ecological disaster. Under the directives of the apex court, the Water and Power Development Authority (WAPDA) was supposed to have installed four water-treatment plants, each with capacity to treat 50 cusecs of contaminated water coming from the MNV Drain before releasing it into Manchar Lake, but WAPDA has so far begun work on just one treatment plant. The cost of each plant is Rs 3.7 billion and the government says it does not have the funds to complete the project. For the past three years, the construction work has remained in its preliminary stage.

The apex court's bench has expressed its displeasure over the 'lack of interest' of the government in a matter affecting the aquatic life, wildlife, and human population of the area. The Supreme Court's interest in the well-being of the lake and the community living in and around it has caused a flurry of activity and interest in the lake in recent years. Perhaps the boat people of Manchar Lake can finally look forward to the prospect of a better future.

Notes

1. Shahid Husain, 'Lake Manchar is Dead', *Down to Earth* (Aug. 2004).
2. Zofeen Ebrahim, 'The Destruction of Pakistan's Manchar Lake', *Thethirdpole.net* (Sept. 2015).
3. Nasir Ali Panhwar, 'Reviving Manchar Lake', *Dawn* (June 2016).
4. 'SC takes notice of lake contamination', *Dawn* (July 2010).

14

Keenjhar Lake: Keeping the Legend Alive

LOCATED AROUND A TWO-HOUR DRIVE FROM KARACHI, Keenjhar Lake is supposed to be a freshwater lake that was formed from two natural lakes in the 1970s as part of the Kotri Barrage Canal Irrigation Project. The idea was to create a reservoir that would supply water to the residents of Karachi. Considered Pakistan's second largest freshwater lake, Keenjhar is located in Thatta district. With its deep blue-green waters and lotus flowers, the lake has been a source of inspiration for poets and writers who hail from the province.

'We want a safe, clean lake. It's not just us but the whole of Karachi that uses water from this lake,' complained a local fisherman when I visited the lake in 2009. Around 900 cusecs of water from the lake are pumped daily into Karachi's water supply. The local fishermen were suffering from a variety of ailments caused by drinking the polluted water of the lake. In each community, around 70 to 80 per cent of the population had some form of waterborne disease. I asked the head of the village near the entrance why his people didn't boil the water: 'It's too much of an effort,' was the reply. 'Besides, boiling the water does not get rid of the diesel it contains!'

I was alarmed to see that trucks, tractors, and construction vehicles were parked virtually within one of the two main entry points to the lake, spewing diesel and exhaust into its water. The drivers were not only washing their vehicles but also cooling themselves off in the lake in the sweltering summer heat.

Around them there were thousands of visitors splashing in the water or swimming in proximity to the shallow shoreline. The lake is huge and the distant shoreline barely visible. They say it is 32 km long and spread over 130 sq. km. Near the shore, pollution was visible: the water was a cloudy grey tint and the stench of diesel assaulted the nostrils, yet the frolicking families seemed to be oblivious of this. 'The lake receives around 10,000 to 15,000 visitors every weekend. As for the trucks, well, there is no one to stop them so the drivers just come right into the water,' said Raheela Memon who was then working with the Indus for All Programme which was attempting to conserve the lake by mobilizing the local community. The Indus for All Programme, which was implemented by WWF-Pakistan from 2007–12, focused on livelihood support and the conservation of natural resources and was in the first phase of a long-term vision for the Indus Ecoregion.

With support from other scientific institutions such as the United Nations Environment Programme, Birdlife International, and the National Geographic Society, WWF carried out the Global 200 analysis in the year 1997, which resulted in the identification of 238 globally important ecoregions. An ecoregion is defined as 'a large unit of land or water harbouring a geographically distinct assemblage of species, natural communities, and environmental conditions'.[1] The Global 200 analysis identified five important ecoregions in Pakistan. The Indus Ecoregion is the only one that lies fully within Pakistan's boundaries while all the others are transboundary. The Indus Ecoregion is one of the 40 biologically most significant ecoregions in the world and the most prioritized of the five Global 200 ecoregions within Pakistan. The ecological significance of the Indus Ecoregion led WWF-Pakistan to concentrate the majority of its efforts on it, and it encapsulates freshwater lakes like Keenjhar.[2]

There are nine villages around the lake: Sonahri, Chill, Ghandri, Chakro, Moldi, Dolatpur, Chilliya, Khambo, and Hillaya with a total population of about 50,000 people. The major language spoken in the area is Sindhi. The community is generally poor and most of them are directly or indirectly dependent on the wetland for their livelihoods. The principal profession is fishing supplemented by agriculture and livestock rearing. The lake also provides irrigation water for crops. The desert plains around the wetland are used for livestock grazing and people also fetch fuelwood from the same area. The local people harvest reeds from the lake which are then used to weave mats and baskets to be sold in the market.[3]

The fishermen were more concerned about their fish catch than the dirty water they were drinking on a daily basis. Over the years, the fish population in the lake has decreased (probably due to the pollution and overharvesting) so they were having a difficult time making ends meet. A total of 55 fish species have been recorded in the lake. Some of these are of high economic value such as the clown knifefish, Indian carp, orange-fin labeo, singhara/giant river catfish, rahu, and common catfish. According to WWF-Pakistan, apart from pollution, the principal threats to the Keenjhar Lake area are overfishing, illegal hunting of small mammals, birds and reptiles, deforestation, and the extensive use of agrochemicals which find their way into the lake through run-off.

How difficult would it be to clean up this lake? I wondered

later as we took a boat ride to the middle of the lake to visit the shrine of Noori. Legend has it that Noori was a beautiful local fisherwoman who lived in a hut on the shores of Keenjhar Lake with her parents. One day a Sindhi king, Jam Tamachi of the Sama dynasty which ruled lower Sindh during the fifteenth century, came to the lake to hunt. He saw Noori's lovely, pure face and instantly fell in love with her. The king actually often went to Keenjhar to hunt, sail, and relax in his royal barge parked on the shores of the lake.

The besotted king married Noori and brought her to his palace to live with him as his queen. A miracle, indeed, but one that has left its physical presence on the lake in the form of an island shrine located where the water is at its deepest: almost 30 ft. The Sindhi poet Shah Abdul Latif Bhittai has woven this romantic tale into his poetry. In his verses he describes how God is pleased with Noori for her humility and submission in all matters. Notwithstanding her great beauty, she was neither proud nor boastful and remained unaffected by the grandeur of Jam Tamachi's palace. The story of Noori and Jam Tamachi is part of the seven folk tales described by Shah Bhittai in his compendium of Sufi poetry, *Shah Jo Risalo*.[4] The lines from the *Risalo*, enacting the trials faced by the seven Sindhi heroines of the folk tales, are sung at Sufi shrines each year throughout Sindh, and especially at Shah Abdul Latif's shrine in Bhit Shah on his death anniversary.

Atop the shrine in Keenjhar Lake lie two graves, one of Noori and the other not of the king (he is buried in the Makli Hills near Thatta) but of some Syed saint. According to one version of this legend, Jam Tamachi's attentions to Noori aroused the jealousy of his other queens who then poisoned her fatally. Noori's grave is today visited by hundreds of tourists on a daily basis. Here, around the shrine, the water is so clear a shade of green that I could actually see the little fish swimming just below the surface.

It is not just diesel that is polluting the lake, which is fed by the Kalri Baghar Feeder Canal that takes off from the Kotri Barrage near Jamshoro. Upstream from Jamshoro, a number of industries on the banks of the river dump their chemical effluents into the Indus. There is no one to stop them from doing so despite the passage of the Pakistan Environmental Protection Act of 1997 which is rarely enforced. It is estimated that only around 1 per cent of industrial wastewater in Pakistan is treated before being dumped into nearby canals and rivers. This is toxic water, teeming with chemicals and persistent carcinogenic organic pollutants. Currently, 47 per cent of Pakistan's population is living without safe drinking water. Hospital records indicate that about 80 per cent of the diseases are either waterborne or airborne. The nation's health bill has increased quite significantly just because the local and provincial governments are not ensuring that the sources of their local water supply are cleaned up.

At Keenjhar Lake, however, all was not doom and gloom. The local fishermen were banding into community-based organizations with the intention of taking ownership of this neglected lake. After all, they had no other source of income and were completely dependent on the lake for their survival. 'If the government helps us, we can kick out the truck drivers and clean up this lake,' they told me. It is a start at least: a clean lake would not only be beneficial for them but also for the entire population of Karachi. As there are hardly any medical facilities nearby, one of the first activities held by the Indus for All Programme was setting up a medical camp. The project staff at the lake held several camps in the villages and treated a large number of patients.

There was also a need to improve the tourist infrastructure. When I visited the lake, tourist facilities were minimal. There were around 80 speedboats but as petrol was expensive in those days few visitors would hire the boats. At any rate, the boats were not equipped with life jackets; in the past there have been quite a number of accidents involving overloaded boats sinking in the middle of the lake. In one such accident a few years ago, twenty-six people died. There are no lifeguards or rescue services in the vicinity of the lake so if a boat gets into trouble, it is on its own.

It is estimated that each week at least two to three people drown in the lake. There is a dire need to post at least two dozen trained lifeguards equipped with fast boats and life-saving equipment. Besides, an emergency treatment centre also needs to be established at the lake to save precious lives.

The Indus for All Programme eventually constructed a proper visitor information centre near the entrance to the lake. The Keenjhar Conservation and Information Centre is now open and running very well, and helps tourists coming from all over Pakistan. It is a very popular place for tourists to visit and has generated some income. The objective of the centre is to provide information and create awareness about nature conservation among both visitors and the local communities. There is a watchtower at the centre providing a wide view of the lake. The centre also offers guided tours, nature camping, birdwatching, educational day trips, and boat safaris. The idea is to promote alternative sources of livelihood to the local communities.

The involvement of the local people was crucial for the Indus for All Programme to succeed. Forming community-based organizations and building up their capacity was a painstaking process and the project staff went about it gradually and steadily. According to Aslam Jarwar, then the site manager at Thatta:

> We told them clearly, whatever you have to do, you do it yourself. We asked them, do you want to solve your own problems? We can only help with the networking. Our emphasis is on self-sufficiency. We told them, this is your environment, your children, your future ... how can you change it?

In this way, the project staff gradually won over the local community. 'We trust WWF,' a fisherman told me, who became an active member of a CBO.

Keenjhar Lake is also a wildlife sanctuary and a Ramsar Site so it is internationally protected under the International Convention on Wetlands, commonly known as the Ramsar Convention. The lake is an important breeding and wintering area for a wide variety of birds. Currently Pakistan has 19 wetlands of international importance, and Keenjhar Lake is one of them. The lake harbours an enormous diversity of migratory birds both in the summers and winters; it was reported to host over 100,000 birds in the 1980s. Today it provides refuge to almost 250 different species of birds: herons, egrets, cormorants, tufted ducks, coots, moorhens, jacanas, and ducks. A total of 14 species visiting the lake are considered scarce, and these include the black-bellied tern, pheasant-tailed jacana, green shank, desert lark, whiskered tern, purple heron, and Pallas' fish eagle. Some other significant species are the greater flamingo, pallid harrier, common kestrel, imperial eagle, and steppe eagle.[5] Keenjhar Lake is also famous for its extensive reed beds, typha grass, and the lotus flowers that grow along its banks.

On our way back after touring the lake, we decided to stop

for cold drinks at the local Pakistan Tourism Development Corporation restaurant (they also run a somewhat unkempt hotel on the lake's shores). The lake clearly needs better restaurant and hotel facilities. Keenjhar Lake was shimmering in the evening sun as we returned by boat. I availed the opportunity and took lots of photographs as we passed the lotus beds near the shore. Shah Abdul Latif Bhittai described the scene in his poetry: 'Upon the waters transparent, along the banks float lotus-flowers, and all the lake rich fragrance showers as sweet as musk when spring-winds blow.' Musing, I thought to myself that if the local people could eventually take ownership of the lake and its management, and the government cooperated with them, Keenjhar Lake could become an ideal tourist destination.

Notes

1. 'What is an ecoregion?', WWF Global. Website: < http://wwf.panda.org/ about_our_earth/ecoregions/about/what_is_an_ecoregion/>
2. 'About Global Ecoregions', *WWF Global*. Website: <http://wwf.panda. org/about_our_earth/ecoregions/about/>
3. 'Keenjhar Lake Brochure', *Foreverindus.org*. Website: <http://foreverindus. org/pdf/sites/keenjhar.pdf>
4. 'Noori Jam Tamachi', *Project Gutenberg Self-Publishing Press (World Heritage Encyclopedia Project, 2016)*. Website: <http://self.gutenberg.org/ articles/eng/Noori_Jam_Tamachi>
5. 'Keenjhar Lake Brochure', *Foreverindus.org*. Website: <http://foreverindus. org/pdf/sites/keenjhar.pdf>

15

Indus Delta: Miracle at Keti Bunder

WHEN I FIRST VISITED KETI BUNDER, A SMALL FISHING TOWN in Thatta district around ten years ago, there was very little vegetation around the ramshackle town overlooking one of the larger creeks. According to one local resident, 'All this was on the banks of the old river. Now it is the shoreline of the sea, and it's washing everything away.' The local people had to purchase drinking water from water tankers that came in from other towns which made the cost of living quite high. Further into the creek lived the poorest of the poor in Pakistan: the fisherfolk inhabiting rickety wooden shacks built on the mud flats of the fan-shaped Indus Delta where the Indus River meets the Arabian Sea.

We had come as a group of environmentalists to visit the surviving Mangrove forests of Keti Bunder, which are categorized as 'protected forests'. The land, lakes, and mud flats had been designated as a Wildlife Sanctuary by the Sindh government. The Keti Bunder union council comprises forty-two settlements of which twenty-eight had already been engulfed by the intruding sea. There are four major creeks in the area: Chann, Hajamro, Khobar, and Kangri or Turchan.[1] There had been a substantial migration to Karachi in recent years where the migrants ended up living in the city's vast slums.

Hajamro and Chann creeks, the two main water channels with small settlements which we visited on that particular trip, were particularly vulnerable because they were losing mangrove cover on a daily basis due to the intense wave action of the advancing sea. We were told that the mud flats in this creek were fast eroding.

Already, given the declining levels of freshwater in the Indus River (in the future, climate change will also adversely affect river flows), the sea had intruded 54 km upstream along the main course of the Indus River. The Indus Delta once used to be home to dense Mangrove forests stretching from Karachi to the Rann of Kutch. Now only some pockets of forest remain. Mangroves need a combination of freshwater and seawater to flourish. Today, the remaining mangroves are under extreme pressure of continuing seawater intrusion, lack of freshwater in the delta, pollution, cutting of wood for fuel, and grazing and trampling by camels.

I still remember my first boat ride along the creek near Keti Bunder town. There was not a patch of greenery, just relentless seawater, the gulls above us, and the bare mud flats. We could even see sand dune formation on the banks of the creek. We all agreed it was an ecological cemetery, and the grinding poverty and despair of the local people living on the mud flats overwhelmed us. 'There is no freshwater, how are we going to survive?' they asked. 'They must release more water below the Kotri Barrage.' The seawater that had intruded into the delta had destroyed thousands of hectares of fertile land and contaminated underground water channels. The government needed to release at least 10 MAF of water each year below the Kotri Barrage (the last barrage on the Indus River) in keeping with the 1991 Indus Water Accords. This was not, however, being done. The Indus River supports one of the world's largest irrigation canal systems which sustain millions of people in the upper area of the Indus Basin, but at the expense of those living downstream in the delta.

These people were living in their wooden huts without electricity or piped water, dependent on marine fishing for their livelihood. WWF-Pakistan was working in the area (they had an office in Keti Bunder town). I for one was sceptical about their prospects of success, particularly when it came to their plans for the rehabilitation of the Mangrove forests. As I was leaving, I wondered how anything could survive here much longer.

I could barely believe the stories related by the local people about how Keti Bunder had once been a prosperous town (the municipality of Keti Bunder even gave loans to the Karachi municipality over a century ago). The people of Keti Bunder would grow red rice, bananas, coconuts, and melons. Now their agricultural lands had either been swept away by the sea or ruined by waterlogging and salinity. The area used to be home to 8 species of mangroves though only 4 have survived. Amongst these, the *Avicennia marina, Aegiceras corniculata,* and *Rhizophora mucronata* can sustain well in saline water. The gradual decrease in freshwater and an increase in saline water has seriously constrained mangrove growth.[2]

What a surprise then to revisit the area five years later in 2009 and discover that the work done by WWF-Pakistan's Indus for All Programme had actually transformed the lives of the people living in Keti Bunder. On my second boat ride into the delta, I noted new plantations of mangroves in and around the villages in the creeks, some right next to the wooden homes of the villagers. Where there was once nothing but mud, I saw mangrove saplings re-emerging (they had planted *Avicennia marina,* the more salt-tolerant mangrove species). I noted wind turbines generating electricity, and water tanks in most of the villages. Above all, the people looked happier and healthier. 'We have cold storage facilities now and we can store our fish and it won't get spoilt. We have drinking water and the wind turbines are good because they give us free electricity,' they told me as I toured one of the villages in Hajamro Creek.

WWF-Pakistan's Indus for All Programme was a six years programme from 2007–12 and one of its principal objectives was to ensure better natural resource management in the Indus Delta, which would contribute to improved livelihoods and sustainable development. Hajamro and Chann creeks were a part of the Indus for All Programme's priority site in Keti Bunder. After Cyclone Yemyin hit the area in June 2007, several rehabilitation projects were carried out by the programme: a boat water tanker was given to the community, fixed potable water tanks were installed in various locations, and thatched huts were especially designed and built for the local communities (priority was given to widows and orphans). The programme also repaired damaged boats belonging to the fishermen. Most importantly perhaps, the programme helped the local community form five community-based organizations to address their own particular problems.

At the time of my second visit, a CBO based in Keti Bunder town was running the water tanker boat service which distributed 4,000 litres of drinking water to each village in the area twice a month. The water was brought from a nearby canal and then stored in Keti Bunder town, from where the boat would replenish its supplies and then make its journey into the creeks. Each community paid for the water (saving on transportation costs), and the women of the village oversaw its fair distribution. The programme also installed solar energy units in schools and mosques, and set up five wind turbines which generated sufficient electricity (for energy-saver bulbs) for around twenty to twenty-five households in the villages where electricity was not available. This resulted in increased economic activity at night, such as the grading of shrimp, and enabled the women to perform their chores more easily. There was also greater social interaction in the evenings.

The CBOs had also been sensitized about the importance of protecting the mangroves by controlling cutting and grazing. Local festivals were organized by the programme in which theatre was used to educate people about the importance of protecting the mangroves. Even the religious leaders of the area had been mobilized to give sermons on nature conservation. The local community was also involved in planting yet more mangroves on mud flats in the creeks.

In 2011, WWF-Pakistan initiated yet another project called Building Capacity on Climate Change Adaptation in Coastal Areas of Pakistan (CCAP) in the delta area. Twenty-five villages in Keti Bunder and thirty-five in Kharo Chann were selected for pilot projects. More solar panels to improve living standards in the coastal areas were introduced in people's homes by the programme. Crab-fattening ponds were also introduced, enabling the local people to safely breed baby crabs which fetch high prices in the markets once they mature.

Around 7,500 ha of mangroves had been planted by WWF-Pakistan earlier with the help of local communities under the Indus for All Programme and another 525 ha were planted through the CCAP project. *Avicennia marina* was mostly planted as it requires less freshwater.

I returned to Keti Bunder yet again in 2014 to look at the new CCAP project. This time around, I noted that soil erosion was becoming a huge problem. Fisherman Sammar Dablo, who had shifted to a new home on a mud flat in Phirt village in Keti Bunder after soil erosion had threatened his prior home, told me that one of the first things he did was plant mangroves around his hut 'to protect the shoreline'. The shrimps and crabs that lay their eggs in the mangrove roots are also beneficial for the villagers of Phirt as they can sell them to markets in the megacity of Karachi just a three-hour drive away.

Due to all the land erosion caused by seawater intrusion, Sammar Dablo's family had migrated three years earlier to their present location along with forty-one other households. The new settlement is located around 2 km inland by boat from their old village, which was on another mud flat facing the mouth of the delta with the Arabian Sea beyond:

> Our old location would be hit directly by the waves from the sea and the winds too were very strong there. We feel much safer here as this place is higher and we have all planted mangroves on this mud flat. It takes around three to four years for the mangroves to mature and become around 5 ft high, depending upon the quality of the soil.

The main problem they face when it comes to mangrove plantation in their area is the threat from camels which come there regularly and eat up the mangrove saplings. 'We try to stop the camel herders from coming here with their camels but they don't listen; we often have to fight with them to chase them off,' Sammar Dablo told me. Indeed, on my boat ride to his village, I passed by several mud flats where camels were freely grazing on the young mangrove plantation.

In the village of Meero Dablo, which is located further inland next to Keti Bunder town, the local community has been planting mangroves themselves for a few years. According to Amina who received training on mangrove plantation from

WWF-Pakistan, 'The mangrove roots are where the fish lay their eggs which helps the fishermen. The mangroves also protect the bund around the village which keeps out the tidal waves and storm surges.'

The good news from the Indus Delta is that WWF-Pakistan's efforts have paid off. In December 2012, a hazard delta-wide mapping study was conducted in the area.[3] This covered the land cover/land use changes (forest changes, land erosion, etc.) utilizing satellite images and geospatial technology undertaken by the Geographical Information System (GIS) laboratory at WWF-Pakistan. The results revealed a significant expansion in the forest area of all the three project sites, i.e. Keti Bunder, Kharo Chann, and Jiwani, where WWF-Pakistan is working. In Keti Bunder, the increase in the forest cover clearly seems to be the result of plantation activities.

The CCAP project also introduced elevated platforms to build raised homes which can protect the villagers from storm surges and tidal flooding. Sakina Ismail showed me her new elevated home in Siddique Dablo village located in Hajamro Creek. It was a neat, wooden home built on stilts. We climbed up a narrow ladder to peek inside. 'Now when the water rises we can save all our bedding, clothes, and utensils and bring our children up here. Everyone in the village wants one of these raised homes, but of course they are expensive to make because of all the wood that is necessary!' she told me.

Notwithstanding these interventions, migration will probably continue in the Indus Delta where the sea level continues to steadily rise. For those living on the mud flats of the delta, on the frontline of climate change, every day is a battle of survival against the seawater. However, with the help of WWF-Pakistan, the people of Keti Bunder are doing their utmost to adapt to climate change, giving not only themselves but others in the delta area hope of a more sustainable future.

Notes

1. 'Keti Bunder Brochure', *Foreverindus.org*. Website: <http://foreverindus.org/pdf/sites/ketibunder.pdf>
2. 'Mangroves to the Rescue', *WWF-Pakistan*. Website: <https://www.wwfpak.org/ccap/pdf/casestudies/Mangroves%20to%20the%20rescue.pdf>
3. 'Delta Wide Hazard Mapping Study—A Case of Keti Bunder, Kharo Chann and Jiwani', *GIS Laboratory,WWF-Pakistan* (Dec. 2012).

16

Daran: Treasure of a Turtle Beach

I FIRST VISITED DARAN BEACH NEAR JIWANI TOWN IN Balochistan, one of the most important marine turtle nesting sites on the coast of Pakistan, in 2005. I remember it as a pristine beach with white sands and clear blue waters. We had to catch a flight from Karachi to Gwadar and then travel by jeep across a rough track to Jiwani because the Makran Coastal Highway was still under construction. On my second visit five years later, I was happy to see that the beach was as picturesque as I remembered it and that the turtles were still arriving and safely laying their eggs. Daran beach is located on the Makran Coast near Iran and—with its towering cliffs, clear waters, and not a plastic bag in sight—it is one of the best beaches I have ever seen.

WWF-Pakistan has been working over here since 1995 to conserve green turtles, an endangered species. Green turtles like to nest on sandy beaches at the foot of cliffs on the shoreline. The cliffs are 1–2 m high in the west and rise up to 30–40 m in the east. The beaches at the foot of the cliffs comprise soft sand and are gently sloping. Daran village is the only settlement in the area and is perhaps why the beach has retained its sanctity.[1]

The turtles come in their hundreds during the hatching season (which begins in August), lumbering out of the water and on to the sand to lay their eggs, digging deep to protect them. Green turtles are usually black-brown or greenish-yellow in colour. They can grow from 100–150 cm and weigh from 110–200 kg. A mature turtle (they can live to be over a hundred years old!) can lay around 70 to 150 eggs in a single season. The turtles nest until March with peak nesting occurring between August and November.

After two months, the hatchlings are born and emerge from the sand pits, falteringly making their way towards the open sea as if guided by some internal radar. It is a remarkable sight: I saw for myself a tiny baby turtle, freshly out of its eggshell, taking its first halting steps towards the vast blue waters just beyond, waving its flippers as a big wave caught it and swept it into the sea. I was shocked to learn that only 1 in 1,000 of these baby turtles actually live to adulthood. Approximately 10,000 are launched each year from this beach and only around 100 actually survive to themselves mate. There are jackals on the beach hunting for the baby turtles and then there are the seagulls just waiting to swoop down and carry them off as they emerge from the sand. Then, even if they do make it to the sea, they might be eaten by a big fish or caught in the large nets of the trawlers that are illegally prowling Balochistan's coastline.

As I nervously watched another one of the hatchlings crawl towards the sea, I decided to take no chances and scooped it up and helped it into the water. As there is a shelf beneath the shoreline here, it is possible to walk quite far into the sea. I walked into the waves, the soft sand cushioning my feet and the clear blue waters lapping at my knees, with the baby turtle in my hands. I eventually let go of the turtle and watched as it swam furiously away into the gentle Arabian Sea. I hoped that this one would somehow survive and live a long life.

Jiwani is located at one edge of Gwadar Bay, one of the

important trading links between Iran and Pakistan. Fishermen use speedboats for fishing and for the transportation of goods and oil between the two countries. These speedboats have increased in number in recent years and, according to WWF-Pakistan, hundreds of them move around the sea near Jiwani. Close to Daran beach there is an area where turtles have been regularly observed mating, but the increase in the number of boats has had a deleterious impact.

It was noted that the oil spills from the boats carrying oil from Iran is a threat to the marine turtles. The use of plastic fishing nets, common in this area, and abandoned nets can have an adverse impact on the turtle populations of the area. Though the turtles are not consumed locally, there has been a history of the utilization of turtle eggs by the local people for the cure of rheumatism. They also used them to treat sick goats and sometimes camels. However, now that the local community has been involved in the conservation of marine turtles, they do not allow people to collect the eggs or damage the nests.

Since 1999–2000, WWF-Pakistan had a site manager for Jiwani, and Daran beach was part of his assignment. He interacted with the local community and was involved in raising awareness about turtles. He motivated the people of Daran to collectively support and protect these once abundant species of the Arabian Sea. The Pakistan Wetlands Programme, implemented by WWF-Pakistan, later formed a local governance structure which was called the Village Conservation Committee (VCC). The VCC comprised community members and local activists, and through awareness campaigns the programme sensitized local communities about the threats being faced by the green turtle population. The VCC was then transformed into the Daran

Conservation Society to ensure its sustainability, and the community participated in a range of activities for the turtle conservation programme. They were trained as turtle watchers and also helped maintain a record of nested eggs and the safe release of hatchlings into the sea.

Community guards were assigned to patrol the beach and chase away jackals at night as most of the laying and hatching takes place at that time. The guards would place wire mesh cages around the turtle eggs to ensure no predator came near them and they would also keep a close watch on any visitors to the beach who might pose a threat to the nests. From 1999–2008, the daily monitoring at night, which took place throughout the year, was undertaken by a team consisting of two members patrolling a 1.5 km stretch of the beach. Observations were made until shortly before sunrise. Data was collected on both the nesting turtles and on the turtles returning to the sea without laying eggs. The local people were of course given training courses on turtle identification, turtle nesting processes, and the transfer of eggs to a hatchery.[2]

For all their hard work over the years, the community was awarded a Living Planet award by WWF-International. 'There are now four beaches that we patrol here: Daran Tak, Shaheed Tak, Deedlo Tak, and Picnic Point,' said Abdul Rashid, who had been working as a guard at Daran and other beaches for over a decade. The beaches are actually located on community land. 'Altogether there are six of us who are paid by the project to protect the turtles. During nesting season, even the children join in to help release the hatchlings safely. The children here are all trained now in protecting the turtles,' he said, pointing to a group of young children who had followed us to the beach to satisfy their curiosity.

According to WWF-Pakistan data, from October 2007

through April 2011, a total of 2,580 nests were protected from which about 26,000 hatchlings were safely released into the sea. The clutch size varied between 78 to 120 hatchlings per nest. The incubation period ranged between 55 to 104 days, depending upon the temperature.[3]

Experts say that the population of green turtles coming to lay eggs on Daran beach has recently been decreasing which may be attributed to the increase in the number of fishing boats. Speedboats are known to strike green turtles swimming on the surface of the sea. Jackals, foxes, and feral dogs are the principal predators of turtle eggs, while seagulls and ghost crabs are the principal predators of hatchlings. Despite WWF-Pakistan having conducted several studies on turtle nesting, a number of facts about the life history of green turtles are still not known. More intensive studies are therefore necessary to help us fully understand this unique species, which often migrate long distances between their feeding grounds and the beaches where they are hatched.

To learn more about the green turtles' migratory patterns and their movements, the Pakistan Wetlands Programme installed thirteen satellite trackers on thirteen green turtles. The satellite tracking revealed that there is daily movement of marine turtles between Jiwani and the Iranian Coast. The most visited sites were Jiwani and Bandar Abbas, with turtles remaining for about 1.5 to 2 months in these areas. Two turtles, one tagged on Astola Island and one at Daran, travelled as far as the UAE and appeared near Umm Al Quwain. The westward movements of the turtles were successfully tracked to Iran, Qatar, and the UAE. The eastward movements of two tagged turtles were tracked to the east coast of India. The turtles travelled along the Makran Coast and reached the Sindh Coast in Karachi, from where they travelled to the east coast of India.[4]

The green turtle is the principal species found nesting at Daran beach; nests of olive ridley turtles are occasionally sighted. Dead hawksbill and leatherback turtles have been washed up by the waves but there are no records of nesting. All species of marine turtles have been categorized as endangered in the IUCN Red List of Threatened Species (2008) and have been declared legally protected under the Balochistan Wildlife Protection Act 1975. The Jiwani Coastal Wetland is also a designated Ramsar Site (Wetland of International Importance) of Pakistan.

There is, however, no legal protection for the marine turtle habitat in Jiwani. WWF-Pakistan has proposed to the Government of Balochistan that they declare it a 'Protected' area.

Aside from their turtle conservation activities, the Pakistan Wetlands Programme also helped to upgrade the local primary school into a two-room structure with a verandah. Around seventeen children from the community had been enrolled there when I visited the school. However, the teacher who had been provided by the government was missing. 'Because it is such a remote place (Jiwani town is around 30 minutes away by road) the government teachers keep running away,' said Abdul Rashid. The schoolchildren had also been involved in beach cleaning activities and awareness-raising walks; and the community regularly celebrated environment days.

Green turtles here have become a part of the children's lives and education, and Abdul Rashid said that 'they have now become a part of our family'. In 2008, the Pakistan Wetlands Programme installed a hybrid wind turbine and solar panel system in the village to provide the local community with electricity. The system generated sufficient electricity to power light bulbs and wireless telephones. The villagers also own some agricultural land nearby where they grow cereals and watermelons.

Abdul Rashid later showed me a record book he kept in his house in the village. It contained seasonal records of nests and hatchlings. His 12-year-old son, Sameer, joined us and told me how he enjoyed spending entire nights helping the turtle hatchlings reach the sea. 'I've been doing this ever since early childhood,' he said smiling shyly. The community-based watch and ward system had certainly protected the nests and helped the turtles survive in the face of enormous odds.

When the Pakistan Wetlands Programme concluded in 2012, the turtle conservation programme also came to an end. There is, however, an urgent need to continue the programme at Daran beach given the observation that turtle nesting has considerably dropped in the area. WWF-Pakistan says it has plans to focus on the beaching habits of green turtles, not only in Jiwani but along the entire coastline. Now, because of all the developmental activities planned in the area under the China-Pakistan Economic Corridor, it has to be ensured that this special beach which is famous for its nesting grounds of green turtles is not endangered by the construction of hotels and other infrastructure in the area.

Notes

1. Muhammad Moazzam Khan, 'Daran-Jiwani Area', submission to the *Convention on Biological Diversity*. Website: <https://www.cbd.int/doc/meetings/mar/ebsaws-2015-02/other/ebsaws-2015-02-template-pakistan-02-en.pdf>
2. Ahmad Khan, 'Pakistan wetlands programme's marine turtle

conservation efforts on Daran Beach, Jiwani, Pakistan', in *Indian Ocean Turtle Newsletter*, no. 17 (Jan. 2013).

3. Umer Waqas et al., 'Conservation of Green Turtle (Chelonia mydas) at Daran Beach, Jiwani, Balochistan', *Pakistan Journal of Zoology*, vol. 43, no. 1 (2011).

4. 'WWF-Pakistan seeks community support for conservation of endangered sea turtles', *WWF-Pakistan* (June 2016). Website: <http://www.wwfpak.org/newsroom/15616_seaturtle.php>

Acknowledgements

I WOULD LIKE TO THANK WWF-PAKISTAN'S SCIENTIFIC Committee members for believing in the concept of my book and giving me a small grant to start and complete the book. I would particularly like to thank Dr Ejaz Ahmad, who has now retired as a senior director from WWF-Pakistan, for editing each chapter and giving me valuable feedback.

Some of the stories that appear in the book are based on my travels for an earlier book, *Green Pioneers: Stories from the Grassroots*, which I co-authored around fifteen years ago for the Global Environment Facility's Small Grants Programme (GEF-SGP) under the auspices of the United Nations Development Programme-Pakistan (UNDP-Pakistan). I would like to thank Fayyaz Baqir, then head of the GEF-SGP for always encouraging my writing on environment and development issues. I am grateful to the support extended by both UNDP-Pakistan and WWF-Pakistan in arranging many of my field visits. I particularly owe many thanks to Rab Nawaz, who still works at WWF-Pakistan as a senior director, and Richard Garstang, former conservation advisor to WWF-Pakistan, and Saleemullah, former head of the UNDP's Small Grants Programme to Promote Tropical Forests, for all their guidance and help in sending me to the field to report on their projects.

I would also like to thank LEAD-Pakistan for providing

me with invaluable training on environment and development that added much needed scientific input into my writing the stories included in the book. My journalistic writing skills were actually honed earlier during my years as a features editor at Pakistan's first independent English news weekly, *The Friday Times*. I am indebted to my editors at *TFT*, Najam Sethi and Jugnu Mohsin, for supporting me to become the writer I am today. Many of the stories that appear in this book were originally feature articles written for *TFT*. Others were written for *DAWN* newspaper, for their weekly magazine section. It was the late Ardeshir Cowasjee, a pioneering environmentalist in Pakistan himself, who encouraged me to write for *DAWN* on environment and climate change issues.

Lastly, I would like to thank my family. My husband, Syed Adnan Haider, for always encouraging me and ensuring that I stay at home long enough to focus on writing this book which is a collection of my travels undertaken in my 20s and 30s and my late father Mohd. Saeed Khan for inspiring me to write this book. If he had not taken me up to the mountains every summer throughout my childhood, I might never have become an environmentalist.

Index

Locations

People

Species